Health Claims and Food Labelling

Food Chemistry, Function and Analysis

Series editors:
Gary Williamson, *University of Leeds, UK*
Alejandro G. Marangoni, *University of Guelph, Canada*
Juliet A. Gerrard, *University of Auckland, New Zealand*

Titles in the series:
1: Food Biosensors
2: Sensing Techniques for Food Safety and Quality Control
3: Edible Oil Structuring: Concepts, Methods and Applications
4: Food Irradiation Technologies: Concepts, Applications and Outcomes
5: Non-extractable Polyphenols and Carotenoids: Importance in Human Nutrition and Health
6: Cereal Grain-based Functional Foods: Carbohydrate and Phytochemical Components
7: Steviol Glycosides: Cultivation, Processing, Analysis and Applications in Food
8: Legumes: Nutritional Quality, Processing and Potential Health Benefits
9: Tomato Chemistry, Industrial Processing and Product Development
10: Food Contact Materials Analysis: Mass Spectrometry Techniques
11: Vitamin E: Chemistry and Nutritional Benefits
12: Anthocyanins from Natural Sources: Exploiting Targeted Delivery for Improved Health
13: Carotenoid Esters in Foods: Physical, Chemical and Biological Properties
14: Eggs as Functional Foods and Nutraceuticals for Human Health
15: Rapid Antibody-based Technologies in Food Analysis
16: DNA Techniques to Verify Food Authenticity: Applications in Food Fraud
17: Advanced Gas Chromatography in Food Analysis
18: Handbook of Food Structure Development
19: Mitigating Contamination from Food Processing
20: Biogenic Amines in Food: Analysis, Occurrence and Toxicity
21: Nutrition and Cancer Prevention: From Molecular Mechanisms to Dietary Recommendations
22: Health Claims and Food Labelling

How to obtain future titles on publication:
A standing order plan is available for this series. A standing order will bring delivery of each new volume immediately on publication.

For further information please contact:
Book Sales Department, Royal Society of Chemistry, Thomas Graham House, Science Park, Milton Road, Cambridge, CB4 0WF, UK
Telephone: +44 (0)1223 420066, Fax: +44 (0)1223 420247,
Email: booksales@rsc.org
Visit our website at www.rsc.org/books

Health Claims and Food Labelling

Edited by

Siân Astley
EuroFIR AISBL, BE
Email: sbastley@gmail.com

ROYAL SOCIETY OF CHEMISTRY

Food Chemistry, Function and Analysis No. 22

Print ISBN: 978-1-78801-073-3
PDF ISBN: 978-1-78801-303-1
EPUB ISBN: 978-1-83916-087-5
Print ISSN: 2398-0656
Electronic ISSN: 2398-0664

A catalogue record for this book is available from the British Library

The Royal Society of Chemistry is a charity, registered in England and Wales, Number 207890, and a company incorporated in England by Royal Charter (Registered No. RC000524), registered office: Burlington House, Piccadilly, London W1J 0BA, UK, Telephone: +44 (0) 20 7437 8656.

Visit our website at www.rsc.org/books

Printed in the United Kingdom by CPI Group (UK) Ltd, Croydon, CR0 4YY, UK

Preface

In June 2016, the EC-funded Framework 7 project BACCHUS (Beneficial effects of bioactive compounds in humans, grant agreement no. 3120900) hosted a workshop in Brussels (Belgium) exploring best practice in health claims and human studies underpinning health claims. The wider objective of BACCHUS had been to develop tools and resources that could facilitate generation of robust and exploitable scientific evidence supporting claims for cause-and-effect relationships between bioactive (food) peptides and polyphenols, and beneficial physiological effects related to cardiovascular health in humans.

To achieve this, BACCHUS assembled 12 leading research and technological centres and 16 small- and medium-sized enterprises (SMEs), including some developing food products that might be suitable for a health claim. Activities were aligned with key aspects of the European Food Safety Authority (EFSA) health claim evaluation process (*e.g.* legislation and dossiers, product/bioactive characterisation, habitual intakes in a population, bioavailability, mechanisms and biomarkers, clinical trials evidence of health benefit) and, as well as new scientific results, user-friendly best practice guidelines were published to support both SME food businesses and research organisations pursuing a health claim. The final workshop presented the BACCHUS outputs and sought feedback from food businesses as well as comparing the European approach with the legislative environment in New Zealand, which had recently undergone significant change with respect to health claims.

This book arose from the workshop, specifically the programme – reflecting the major interests of the various sectors – and feedback received from participants, particularly how the European food industry, dominated by SMEs, might find their way through such a technical process successfully. It is not a comprehensive exploration of the various legislative frameworks

Food Chemistry, Function and Analysis No. 22
Health Claims and Food Labelling
Edited by Siân Astley

Published by the Royal Society of Chemistry, www.rsc.org

around the globe that consider food labelling (nutrients and allergens) and health claims, but rather a readily accessible starting point for researchers and food businesses alike, providing signposts to and around the enormous body of literature and legislation covering these topics.

The main focus is on Europe, with **Chapter 1: European Food Information Regulations: Recent Developments** considering the most recent changes, specifically mandatory labelling of nutrients through the Regulation on the Provision of Food Information to Consumers (FIC) (1169/2011/EC), and the history behind this Regulation and different labelling requirements. **Chapter 2: Impact of Health Claims Regulations on the Food Industry and Future Innovation Potential** looks at the Nutrition and Health Claims Regulation (NHCR, EC No 1924/2006) more specifically, and how economic competitiveness, product innovation and better understanding amongst consumers can be achieved for mutual benefit. **Chapter 3: Supporting the European Food Industry to Meet Nutrient Labelling Regulations** focuses on the needs of food businesses and how information provided, and actions undertaken by competent authorities, trade associations and others across the EU or within EU Member States, can best inform businesses of their responsibilities, including access to legislation and the availability of guidelines and training.

Chapter 4: Health Claims Regulation: Opportunities and Challenges in Europe explores how claims about nutritional and health properties of foods and beverages are regulated in Europe, under Regulation No 1924/2006, and the detailed processes in place to ensure scientific validity as well as how, over time, difficulties with this legislation have emerged including monitoring of compliance, as managed at the national (EU MS) level. Such constraints have, arguably, limited the number of claims submitted to EFSA for scrutiny and might disadvantage disproportionately the smaller food business, which make up the majority of the European food and beverage sector.

Chapter 5: Use of Nutrition and Health Claims and Symbols on Prepacked Foods in Europe: From Consumer Exposure to Public Health Implications reports on the use of nutrition and health claims in the market, as reported by the pan-European CLYMBOL. Whilst CLYMBOL found that foods labelled with nutrition and/or (specific) health claims, for the most part, used the mandatory nutrition declaration, a quarter of the foods with nutrition claims did not include the content of the nutrient/substance to which the claim referred or did not meet defined conditions for an authorised claim and, in some cases, unauthorised claims were used. Generally, the occurrence of (non-specific) health claims was lower than nutrition claims, possibly because these claims were not recognised as health claims, which is in itself a major issue. Related to this, **Chapter 6: Measuring the Effects of Health Claims and Symbols: The CLYMBOL Methodological Toolbox** describes a tool designed to provide guidance for policy-makers and food businesses who need simple and robust, but scientifically sound, methods to test consumer understanding of claims and symbols, and obtain insights into potential effects on purchasing and consumption.

In contrast, **Chapter 7: Food Allergen Labelling Regulation** considers the mandatory indication of priority allergens. Food allergies have risen in prominence over the last 20–30 years and, currently, as there is no accepted cure, individuals have to practise life-long avoidance of problem food(s). Although rare, individuals with food allergies can be hospitalised as a consequence of accidental exposure and there is risk of a fatal reaction (anaphylaxis). Thus, allergic consumers need to identify, without doubt, products that are safe to eat and those that must be avoided. Despite the introduction of mandatory priority allergen labelling, there are significant issues with so-called precautionary labelling, where consumers are warned about a potential risk arising from cross-contamination. **Chapter 8: The Impact of "Labelling" on the Beliefs, Attitudes and Behaviours of Consumers with Food Allergy: A Multilevel Perspective** reviews the impact of increasingly widespread, and sometime indiscriminate, use of precautionary labelling on food-allergic individuals or their carers, and healthcare professionals, and how the existing approaches might be revised, based on emerging scientific knowledge, to provide greater protection for consumers and the food industry.

In contrast, **Chapter 9: Food Labelling Practices in North America: Canada and the USA; Chapter 10: Supporting Food Health Claims in New Zealand's Regulatory Environment** and **Chapter 11: Latin American Food Labelling: Review and Recent Developments** describe labelling approaches in two regions (North and South America) and one country, New Zealand, reflecting the original comparison first made in BACCHUS.

New Zealand has brought together food scientists, regulatory considerations and consumer perspectives and provided tools for food businesses making health claims and improve communication whilst Canada and the USA have led the way, setting policy rationales as means for adopting food labelling regulatory requirements, with mandatory declarations for nutrients and priority allergens. In both the USA and Canada, regulators have also set conditions for the discretionary use of statements such as health claims, which are different from those prescribed in Europe. In contrast to North America, Europe and New Zealand, Latin American countries are characterised by all forms of malnutrition as well as a significant burden of non-communicable diseases, such as obesity and diabetes. There is also increased consumption of ultra-processed foods, high in sugars, saturated and trans fats and sodium. To date, three countries have implemented mandatory front-of-package labelling, as a tool for improving diets and communicating potential health benefits, and tackle both under- and over-nutrition.

Siân Astley

Contents

Chapter 1 European Food Information Regulations: Recent Developments **1**
Stephen Pugh

1.1 Introduction 1
1.2 Why Nutrition Information? 3
1.3 Codex Alimentarius 4
1.4 Current Labelling Requirements 5
1.5 Trans Unsaturated Fatty Acids (Trans Fats) 6
1.6 Front-of-Pack (or Traffic Light) Labelling 7
1.7 Nutrition Labelling and Alcoholic Drinks 8
1.8 Tolerances 9
References 11

Chapter 2 Impact of Health Claims Regulations on the Food Industry and Future Innovation Potential **13**
Katia Merten-Lentz

2.1 Introduction 13
2.2 From Food Safety Policy to Nutrition and Health Claim Regulation 14
2.2.1 General Food Law and Business Operators 14
2.2.2 Harmonisation of Nutrition and Health Claim Legislation 14
2.3 Scope of EU Nutrition and Health Claim Legislation 16
2.3.1 Claims 16

Food Chemistry, Function and Analysis No. 22
Health Claims and Food Labelling
Edited by Siân Astley

Published by the Royal Society of Chemistry, www.rsc.org

2.4 From a Particular Characteristic to Nutrition and Health Claims 18
2.4.1 Nutrition Claims 18
2.4.2 Health Claims 18
2.4.3 Generic (Non-specific) Claims 19
2.5 Scientific Substantiation for a Claim 19
2.6 Authoritative Statement of NHC 19
2.6.1 Application 20
2.6.2 From Risk Assessment to Risk Management 20
2.6.3 Nutrient Profiles Issue 21
2.7 Permitted Claims 22
2.7.1 Permitted Nutrition Claims 22
2.7.2 Permitted Health Claims 22
2.7.3 Suspension Claims 22
2.8 Conclusion 23
References 24

Chapter 3 Supporting the European Food Industry to Meet Nutrient Labelling Regulations 28
Marie Macháčková, Mark Roe, Marija Ranic and Paul Finglas

3.1 Introduction 28
3.2 Access to Sources of Food Law 30
3.2.1 European Union 30
3.2.2 EU MS 31
3.3 Guidelines 32
3.3.1 Guidelines: Formats 32
3.3.2 Guidelines on Obtaining Nutrient Values 44
3.4 Training Courses 50
3.5 Consumer Education 51
3.6 Conclusion 51
Acknowledgements 56
References 56

Chapter 4 Health Claims Regulation: Opportunities and Challenges in Europe 63
Judith Buttriss

4.1 Introduction 63
4.2 What Does the Law Require? 63
4.3 What Types of Claim Are Covered? 64
4.3.1 Nutrition Claims 64
4.4 Health Claims 65

4.5 Article 13.1 Health Claims 65
4.6 Article 13.5 Health Claims 67
4.7 Article 14 Health Claims 67
4.8 Reference to General Non-specific Health Benefits 68
4.9 Nutrient Profiles 68
4.10 Regulation No 1924/2006 Evaluation: Process in Outline 69
4.11 Novel Foods Legislation 71
4.12 Consumer Understanding of Health Claims 73
4.13 Impact of the Regulation on Small Businesses 74
4.14 Conclusions 75
Acknowledgements 76
References 76

Chapter 5 Use of Nutrition and Health Claims and Symbols on Prepacked Foods in Europe: From Consumer Exposure to Public Health Implications 79
Igor Pravst, Anita Kušar, Krista Miklavec, Sophie Hieke, Monique Raats and Mike Rayner

5.1 Introduction 79
5.1.1 Prevalence of Nutrition and Health-related Claims on Prepacked Foods 81
5.1.2 Nutritional Quality of Foods Labelled with Claims 83
5.1.3 Ensuring Informed Consumer Choices 84
5.2 Methods and Materials 85
5.3 Results and Discussion 86
5.3.1 Labelling with Nutrition Declaration 86
5.3.2 Labelling of Foods Carrying Nutrition Claims 87
5.3.3 Labelling of Foods Carrying Health Claims 87
5.4 Conclusions 89
Acknowledgements 91
References 91

Chapter 6 Measuring the Effects of Health Claims and Symbols: The CLYMBOL Methodological Toolbox 94
Sophie Hieke and Klaus G. Grunert

6.1 Introduction 94
6.2 Criteria to Assess Methods Measuring the Effects of Health Claims and Symbols 95
6.2.1 Theory-driven Criteria 95

6.2.2 General Criteria on Validity and Reliability 96
6.2.3 Feasibility in the Study of the Effects of Health Claims and Symbols 96
6.3 Methods for Testing Consumer Understanding of Health Claims and Symbols 96
6.3.1 Consumer Understanding Test (CUT Method) 97
6.3.2 Soft Laddering Interviews (Face-to-Face) 98
6.3.3 Hard Laddering (Online) 98
6.3.4 Conclusions 99
6.4 Methods for Testing Effects of Health Claims and Symbols on Consumer Purchases 99
6.4.1 Sales Data 100
6.4.2 Eye-tracking Study 100
6.4.3 Electrodermal Reactivity Study 100
6.4.4 Implicit Attitude Study 101
6.4.5 Conclusion 101
6.5 Methods for Testing Consumption Effects of Health Claims and Symbols 102
6.5.1 The Breakfast Study 102
6.5.2 Conclusion 103
6.6 Recommendations for the Use of Methods by Various Stakeholders 104
6.6.1 Recommendations for Regulators 104
6.6.2 Recommendations for the (Food) Industry 104
6.6.3 Recommendations for Consumer Scientists 104
References 105

Chapter 7 Food Allergen Labelling Regulation 107
Luca Bucchini, Matthew Daly and E. N. Clare Mills

7.1 Introduction 107
7.2 Food Allergen Regulation and Labelling 108
7.2.1 Mandatory Allergen Labelling 108
7.2.2 Regulated Priority Food Allergens around the World 110
7.2.3 Allergen Declarations 116
7.2.4 Voluntary Precautionary Allergen Labelling 118
7.3 Allergen-free Food Regulation and Labelling 120
7.4 Conclusions 123
References 123

Chapter 8 The Impact of "Labelling" on the Beliefs, Attitudes and Behaviours of Consumers with Food Allergy: A Multilevel Perspective **127**
Audrey Dunn Galvin

8.1 Introduction 127
8.2 The Individual (Micro) System 129
8.3 The Society (Macro) System 132
8.4 What Can Help From a Multilevel Perspective? 134
References 135

Chapter 9 Food Labelling Practices in North America: Canada and the USA **141**
Samuel Benrejeb Godefroy and Kenneth Teng

9.1 Introduction 141
9.2 Food Labelling Measures Imposed as a Risk Management Approach 142
9.2.1 Nutrition Facts Labelling 143
9.2.2 Mandatory Labelling of Mechanically Tenderised Meat Products 150
9.2.3 Mandatory Food Allergen Labelling Measures 152
9.3 Discretionary Food Labelling Measures Framed with Regulatory Requirements 155
9.3.1 Health Claims 155
9.3.2 Gluten-free/Allergen-free Claims 157
9.4 Conclusion 159
Competing Interests Declaration 159
Contributions 159
References 159

Chapter 10 Supporting Food Health Claims in New Zealand's Regulatory Environment **164**
Carolyn E. Lister

10.1 Introduction 164
10.2 Navigating the Claim Options 166
10.2.1 Nutrient Content Claims 167
10.2.2 Health Claims 167
10.2.3 The New Zealand Food Composition Database – A Useful Tool to Support Claims 169

10.3 Antioxidant Claims 171
10.4 Developing Branded Products: **vital**vegetables® 172
10.5 Developing a Self-substantiation Dossier 173
10.6 Conclusions 174
Acknowledgements 176
References 176

Chapter 11 Latin American Food Labelling: Review and Recent Developments 179
Marcela Reyes and Camila Corvalan

11.1 Background 179
11.2 FOP Labelling in Latin American Countries 183
11.2.1 Mandatory FOP in Latin American Region: Cases of Mexico, Ecuador and Chile 184
11.2.2 Other Countries in the Latin American Region 188
11.3 Conclusions 188
References 189

Subject Index 198

CHAPTER 1

European Food Information Regulations: Recent Developments

STEPHEN PUGH

Sugarwise, 5 Signet Court, Swann Road, Cambridge CB5 8LA, UK
Email: stephen.pugh@eufoodlabellingspecialists.eu

1.1 Introduction

The major change in food labelling in the European Union (EU) is the introduction of mandatory labelling for nutrients of public health interest on prepacked foods. This might not seem like a major change in the UK but, in other EU Member States (MS), the market for prepacked foods is not as well established but these and other markets are growing quickly.

The push for mandatory labelling comes from a multitude of directions. The two main protagonists are MS government health departments and from consumer organisations. Governments, and in particular health departments, are concerned about the rise in non-communicable diseases, such as obesity, and better nutrition labelling is seen as a way of empowering and/or nudging citizens to improve their diet.[1] Consumer organisations are being pushed from two directions: members are concerned about the nutritional content of foods, but also the addition of fat, sugar and salt, which increase bulk or palatability of food products without any nutritional benefit or improved quality.

With food labelling being an issue that affects trade across the EU, and largely considered a European Commission (EC) competence, the EC also has

Food Chemistry, Function and Analysis No. 22
Health Claims and Food Labelling
Edited by Siân Astley

Published by the Royal Society of Chemistry, www.rsc.org

an interest. However, health *per se* is an area where MS have priority, making nutrition labelling a joint competence. However, such has been the pressure from MS to harmonise nutrition labelling across the EU, the EC incorporated rules on previously voluntary nutrition labelling in the new regulation on the provision of food information to consumers (FIC) (1169/2011/EC),[2] making nutritional labelling mandatory and standardised across the EU.

In some areas, nutrition labelling has been mandatory for many years, following the introduction of Nutrition and Health Claims Regulations[3] in 2006, together with fortified foods regulations.[4] If manufacturers make a nutrition or health claim on a food, then a nutrition declaration according to the rules on voluntary nutrition labelling was required. The same applied to fortified foods: if a manufacturer fortified a food, then a nutrition declaration was mandatory. This voluntary nutrition labelling was governed by regulation 90/496/EC,[5] and the provisions for presenting this information have been copied over to FIC with a few small changes, *e.g.* labelling of sodium as salt, the order of nutrients and nutrients that must be included. These changes reflect public health concerns of MS governments and can be regarded as future-proofing the FIC regulation.

Given the lack of nutritional information on prepacked foods within the EU previously, particularly from medium-sized food businesses (50–250 staff, 10–50 M Euro turnover or 10–43 M Euro balance sheet total), regulators initiated a longer lead-in period for mandatory nutrition labelling. Requirements for general labelling came into law on 13 December 2014, but nutrition labelling was not mandatory until December 2016. The aim was to promote compliance with the new FIC regulation by fitting in with the normal labelling cycle for medium-sized businesses (*i.e.* approximately 3 years).

The primary aim of nutrition information is empowerment, *i.e.* knowing the nutritional content of a product enables consumers to make informed choices about their diet. In fact, labelling alone makes little difference to consumer behaviour but, combined with national programmes of education, this information can help consumers adopt a healthier diet. Similarly, nutrition labelling is of limited value without public health information that makes consumers aware of the nutrition contents of foods, what these values mean and how they can be used to eat more healthily. In the EU, on a population basis, nutrients such as sugar, fat and salt – as well as energy intake as a whole – need to be reduced. Whether or not individuals need to amend intakes depends on their circumstances. Having amounts of these nutrients on packaging empowers some consumers to make more informed choices.

Actual intakes, however, depend on how much is eaten, *i.e.* portion size. Food producers make strenuous efforts to determine portion size for their products, but consumption depends on the individual, *e.g.* age, access to foods, appetite. Currently (*i.e.* 2018), there are no defined portion sizes in the EU and, given the complexity of the subject, this is unlikely to change in the near future, despite lobbying from a range of stakeholders across Europe. In the meantime, food manufacturers are able to indicate nutrition content based on portion size as well as per 100 g, and many have chosen to do so for a variety of reasons. In many cases, portion size is obvious (*e.g.* one can of a soft beverage or small packet of potato crisps) but, in many cases, particularly large packages, portion sizes are less intuitive (*e.g.* breakfast cereals) and more controversial (*e.g.* chocolate bar).

1.2 Why Nutrition Information?

With changes in the retail environment and consumer preferences for pre-prepared food products, EU MS governments and consumer organisations have been calling for better information on food labels, and this included nutrition information. These calls began in the late 1980s, which is reflected in the directive on voluntary nutrition information that was published in 1990 and a Codex standard for nutrition information adopted in 1985. Increased rates of obesity throughout the developed world, and also amongst some developing countries as GDP increased, added urgency to these calls for better information.

Worldwide, the prevalence of obesity nearly doubled between 1980 and 2008. More than half of all men and women in the World Health Organization (WHO) European Region are overweight, and about a quarter of women (23%) and 20% of men are obese.[6] Based on latest estimates in EU MS, being overweight affects 30–70% and obesity affects 10–30% of adults. The number of overweight infants and children in the WHO European Region has risen steadily since 1990, meaning more than two-thirds of children who are overweight before puberty will be overweight in early adulthood. Childhood obesity is associated strongly with risk factors for cardiovascular disease, type 2 diabetes, orthopaedic problems, mental health disorders, underachievement in school and lower self-esteem.[6] Recommendations for managing and reducing obesity emphasise the need for cooperation amongst different actors not just individuals and, in addition to increased levels of physical activity, information about foods is key to consumers adopting a healthier diet.[7]

Nutritional poverty or poor diet, where energy intake is in excess overall and the proportions of nutrients imbalanced, particularly with respect to high intakes of saturated fat, salt and sugar and insufficient consumption of fruit, vegetables and fibre, is a risk factor for weight gain and obesity. One way of addressing this is to provide information about the amounts of fat, salt and sugar in foods. However, this information is not always clear (*e.g.* intended portion size) and those at greatest risk of nutritional poverty have limited access to healthier foods (*i.e.* local food businesses are predominantly fast food retailers). Also, understanding public health information, supporting application nutrition information, is highly variable. Thus, some EU MS governments have encouraged manufacturers and retailers to provide interpretations around the nutrition declaration (*e.g.* traffic light indicators) or nutrition facts (*e.g.* dietary reference values and guidelines[8]) to help consumers identify and control intakes of foods high in fat, salt and sugar. However, interpretations are difficult (good or bad for whom, *e.g.* full fat milk – recommended for children but not adults), and some manufacturers and retailers have felt unfairly singled out for criticism, particularly those making or selling products high in sugar, salt or fat (*e.g.* chocolate). Of course, it is precisely these products that consumers need to cut down on or eliminate if they are to have a healthy diet and control their weight.

Formats of nutrition declarations and nutrition facts have their origins in the Codex standard. Some EU MS governments have considered front-of-pack labelling, health markers or other ratings helpful for consumers to interpret information in the nutrition declaration. The very need for these and the plethora of information around interpretation, as well as the public debate of experts around their relative value, demonstrates the complexity of nutrition declarations and nutrition facts for the general public.

1.3 Codex Alimentarius

The genesis of the requirement for mandatory nutrition labelling are guidelines for nutrition labelling under the auspices of the General Standard for the Labelling of Prepacked Foods produced by the Codex Alimentarius Commission. Codex was established by the World Health Organization (WHO) and the Food and Agriculture Organization (FAO) of the United Nations (UN) to develop international food standards for the protection of consumer health and to ensure fair practices in the food trade. The General Standard for Labelling of Prepackaged Foods (CODEX STAN 1-1985)[9] was amongst the first Codex standards to be adopted, together with other general Codex texts (*e.g.* hygiene, contaminants, pesticide residues, food additives), and forms an important pillar of the Codex food standards system.

CODEX STAN 1-1985 was revised and enlarged extensively in 1991 and there have been numerous amendments and additions since then. These changes have ensured the Standard meets the needs of modern consumers. As with all texts of this nature, there is some interpretation around the text and guidelines have been produced to help promote consistency. For nutrition information, guidelines were produced initially, following the introduction of the CODEX STAN 1-1985, but implementation has been left to national governments. The implication is that food information should be provided if a government feels the issue is a concern, *e.g.* food hygiene/safety.

CODEX STAN 1-1985 gives a background to the provision of nutrition information and introduced priority nutrients as well as supplementary nutrients that might be given. FIC (1169/2011/EC[2]) nutrients correspond exactly to these, and CODEX STAN 1-1985 guidelines also allow for a certain amount of tolerance around figures on the label for enforcement, which allows for seasonal and production variations. The guidelines also specify nutrients that should be together on the label along with appropriate wording for each. In most nutrition labelling schemes, there are five nutrients of public health interest, *i.e.* energy, fat, carbohydrate, salt and protein. Some national governments or regional legislation have refined further the categories of fat (saturated, polyunsaturated and monounsaturated) and carbohydrates (sugars). There are also supplemental nutrients that may be included (*e.g.* fibre) or nutrients (*e.g.* vitamins) that must be included if a claim is made.

Within Codex, there are a couple of issues that have not been resolved, specifically trans fats and interpretive nutrition labelling, which have been left with national governments to address as they see fit. Codex guidelines allow for interpretive labelling, but they are complex and difficult to understand. The US Food and Drug Administration (FDA) has tried to simplify the language in their nutrition facts table, but which foods might be better as part of a balanced diet is still open to consumer interpretation. The Codex Committee on Food Labelling (CCFL)[10] has acknowledged some interpretation of nutrition declarations would be helpful, and a project to develop guidelines has started.

1.4 Current Labelling Requirements

EU nutrition labelling rules for prepackaged foods are set out in FIC (1169/2011/EC).[2] The most basic requirements are straightforward: a nutrition declaration is mandatory for (most) prepacked food on retail sale. There are seven nutrients of public health interest – energy, fat, saturated fat, carbohydrates, sugars, protein and salt. Other supplementary nutrients can be added on a voluntary basis, or on a mandatory basis if a claim is made around one of these nutrients. The nutrition declarations must be given on a 100 g or 100 cm^3 (100 mL) basis. However, they can also be given as per portion or serving unit. Significantly, these nutritional values must be on an "as served" basis. The amounts of nutrients in the product can be calculated or measured directly through analysis. Composition data may be taken from established databases (*i.e.* food composition tables).

Supplementary nutrients include mono- and polyunsaturated fats, polyols, starch, fibre and any of the vitamins or minerals listed (Point 1 of Part A of Annex XIII of the FIC), and present in significant amounts.

Amounts of food per portion or per serving is at the discretion of food manufacturers but must be realistic. There is a lot of criticism around overly optimistic portion sizes, with a view to ensuring nutritional values are as low as possible. In fact, portion or serving sizes are very difficult to judge and can have positive or negative effect on sales. It is easier where products are predivided into discrete units (*e.g.* individual cakes), when market forces will decide whether these are too small, and sales suffer regardless.

There are a number of exemptions to nutrition labelling, which revolve around unprocessed foods defined in hygiene regulations (852/2004/EC)[11] or foods of little or no nutritional value (*e.g.* natural mineral waters, tea and coffee). Flavoured teas are also exempt from a nutrition declaration, as long as the flavouring adds no nutritional value to the product; the same does not apply to flavoured coffees. Two other exemptions apply to foods (1) with a largest surface area of less than 25 cm^3 (*e.g.* individual sweets/candies), as it would be difficult to include a nutritional declaration in the minimum font size on products of this size; and (2) food, including handcrafted products, supplied directly by the manufacturer in small quantities to the final consumer or local retail establishments supplying the final consumer. Finally, to help small businesses expanding to a larger marketplace, nutritional labelling only becomes mandatory after these companies have expanded beyond county borders.

1.5 Trans Unsaturated Fatty Acids (Trans Fats)

Trans fats occur naturally in some animal fats, such as beef tallow and buttermilks, but they are also an isomeric by-product of dehydrogenation of saturated fatty acids (converting saturated to unsaturated fatty acids). Unsaturated fatty acids (*e.g.* palmitoleic, oleic, myristoleic, linoleic and arachidonic acids) are added to products to give more desirable properties (*e.g.* mouth feel, ease of processing). Trans fats are also produced from the partial hydrogenation of polyunsaturated fatty acids, which is used to reduce the risk of rancidity, *i.e.* extend shelf life.

During FIC negotiations, issues around trans fats, particularly their link with coronary heart disease, were emerging from across the world. Given scientific recommendations from various expert committees to reduce intakes of trans fats, the FIC Working Group faced a dilemma; requiring trans fats to be declared on labels could have the effect of freezing levels in foods whereas the recommendation was that concentrations of trans fats in foods should be as low as possible. The EC opted not to include trans fats in the list of supplementary nutrients, meaning they could not be included on labels. Then, the EC and MS governments agreed to work with food manufacturers to reduce levels of trans fats in foods, and the EC published a report on these efforts in December 2015. The report gave a series of options that could be adopted to ensure trans fat levels were as low as possible for European consumers. These options were mandatory labelling of trans fat content, legislation to limit the amounts of trans fat in food, voluntary agreements with industry to reduce the amounts of trans fat in food and guidance for setting national legal limits.

In December 2017, the EC published a public consultation on trans fats,[12] giving three options:

- Option 1: Prohibition of the use of trans fats in foods, either through regulation or self-regulation, which refers specifically to industrially produced trans fats.
- Option 2: Labelling of trans fats content, acknowledging that labelling would cover both industrially produced trans fats and those produced by ruminants.
- Option 3: Limits on the amounts of partially hydrogenated trans fat in food, which refers specifically to partially hydrogenated trans fats.

The deadline for responses was 9 February 2018 and the outcomes will be published in due course. However, regardless of any new legislation on trans fats in prepackaged foods, the issue of trans fat in foods sold loose (*i.e.* without labelling) will remain. Surveys of bakery products and biscuits have found high concentrations of trans fats in these products and continued vigilance is required to ensure trans fats from the diet are kept to a minimum.

1.6 Front-of-Pack (or Traffic Light) Labelling

The exact requirements of the nutrition declaration are technical, and difficult to interpret. Furthermore, research has shown that time spent choosing food products is relatively small. Thus, interpretive labelling aims to help consumers opt for healthier food products. There are a number of different schemes globally, but they fall into three categories: (1) analysis of the nutrient composition to determine whether it is healthy or not, (2) graphical representation to highlight the nutritional content, and (3) simplified nutritional information in a more stylised form. The Dutch keyhole system is an example of analysis (1), and falls under the definition of a health claim in the EU. The Australian and New Zealand star system is a graphical representation (2). The traffic light labelling in the UK[13] offers more stylised information (3) through colour coding for high (red), medium (amber) and low intakes for the priority nutrients.

The traffic light labelling scheme has been around for more than a decade and, in its current guise, has been in operation since 2016, when the UK Department of Health (DoH) issued new guidance that took FIC into account as well the needs of UK food retailers. The original scheme was adopted by most UK retailers, but there were important exceptions. The new scheme was adopted universally and well received by UK consumers, but there remains debate over whether or not consumers use or understand the information correctly.

A lot of research has been carried out on the effectiveness of traffic light labelling in the UK. Work carried out by Sacks, Rayner and Swinburn[14] was largely ambivalent about the effectiveness of the scheme. Work published by Sainsbury on their website found that over half of shoppers in their stores thought traffic light labelling was useful when making healthy choices.[15] The UK DoH research was enthusiastic about the scheme[16] but, as stated by Sacks *et al.*[14] highlighted, it is difficult to assess the effectiveness of a passive scheme. Other researchers have pointed out there is merit in having such a scheme, as consumers begin to understand which products are healthier and repeat purchases with reassurance from traffic light labelling information.

For those unfamiliar with traffic light labelling, the portion element of the nutrition declaration is repeated on the front of packs in both kilojoules and kilocalories, as a simple numerical amount. Amounts of fat, saturated fats, sugars and salt are also repeated, as numerical values, but the background of the declaration is coloured according to the amount present in the food. Red is given to the higher levels and green to lower levels, based on recommended daily intakes for an average adult. Values in between are coloured amber. The cut-off for green is defined in both UK guidance on the traffic light labelling scheme and in the health claims regulation (1924/2006/EC).[17] For fat, this is less than 3 g per 100 g of a food; for sugar less than 5 g per 100 g; and for salt less than 0.3 g per 100 g.

Figures ascribed as red are more complex as high levels of fat, saturated fats, sugars and salt nutrients are not defined in any EU regulation or guidelines. Instead, values are defined as high (red) if 100 g of the product "as served" contains 30% or more of the recommended dietary reference value for an adult; 30% was selected based on consumption of three largely energy-equivalent meals per day. This is reduced to 25% for liquids, as liquids can be more readily consumed than solid foods. These figures were put out for consultation prior to application, but not challenged to any great degree.

The scheme as a whole has been criticised and the UK was referred to the EC as having introduced a non-tariff barrier to trade. The argument was that food imported into the UK from a country without a similar scheme would be perceived as unhealthy and, consequently, at a disadvantage in the UK marketplace. However, such an accusation, based on a passive labelling system, is difficult to prove (or disprove): consumers are habitual in purchasing habits and have chosen one product over another, perhaps or not influenced by traffic light labelling, will purchase that product again without remembering what triggered their choice. In 2018, the EC was yet to rule on this matter and, with BREXIT imminent (October 2019 – at the time of publication), the discussion could be rendered mute, subject to new trade agreements.

1.7 Nutrition Labelling and Alcoholic Drinks

Alcoholic drinks are exempt from FIC regulations with respect to ingredients lists and nutrition information, provided they are over 1.2% by alcohol volume. This exemption from additional information on ingredients and nutrition was the subject of much debate prior to adoption of regulation. Most EU MS were in favour of having this additional information and it was more a question of how and when, rather than if. Full ingredients labelling might have led to some anomalies, *e.g.* wine can be listed as a category in the ingredients if it complies with the requirements stipulated in the Common Agricultural Policy (CAP).[18] Thus, a full ingredients list on a bottle of wine would go against the requirements for other fermented products.

The debate around nutrition declarations was also coloured by the desire to control products such as alcopops. However, there is no established definition of an alcopop, even though most involved in the debate understood the meaning of the term, *i.e.* sweet, flavoured alcoholic beverages with relatively low alcohol content (*e.g.* 3–7%), the advertising for which targeted younger people. Some drinks manufacturers also felt unfairly penalised, suggested one sector of the alcoholic drinks industry – specifically wine – was favoured over others. In the absence of a legally established definition, the EC decided a fact-finding exercise on different types of alcoholic drinks and consumer expectations was required and the issues of labelling should wait until more information had been collected.

In the meantime, other issues have become apparent, such as how nutrition declarations should be expressed and, specifically, whether these should be the same for all foods (*i.e.* as a fraction of 100 g of the product as served) or according to serving or unit sizes. These issues are important when considering the nutrition content of a spirit with a beer, *i.e.* 100 g of a spirit (*ca.* 100 mL compared with a single [small] measure of 25 mL or [large] 50 mL) seems excessive compared with 100 g of beer (*ca.* 100 mL compared with third [160 mL], half [250 mL], two-thirds [330 mL] of a pint and multiples thereof). Providing any information for alcoholic drinks is complex, not just in terms of nutritional content and tax, but also health and safety, *i.e.* "drink responsibly" public health messages. Thus, manufacturers have been looking at innovative methods for delivery of this information collectively.

The EC published their report on 13 March 2017.[19] The report gives an excellent background on the issues, although there are some omissions (*i.e.* there is no mention of pre-mixed alcoholic drinks such as gin and tonic) and anomalies. For example, served alone, tonic requires a full ingredients list and a nutrition declaration but, served with gin, it requires neither. In most of the drinks sector, there are EU regulations defining the composition of alcoholic drinks,[20] but this does not apply to beer and ciders. For spirits and wine, the name of the product defines the composition and listing the ingredients would be relatively simple (*i.e.* for a bottle of wine, the ingredients would be wine; for a bottle of whisky, it would be whisky). However, the same cannot be said for a bottle of beer, where the ingredients would be water, malt, hops and yeast. Regardless, a requirement to declare the ingredients would not be too onerous and would inform consumers, enabling them to make choices about the product they drink.

For nutrition information, energy content of the alcoholic drink is the most important nutrient. Most, if not all, of this comes from the alcohol and it is a fairly simple calculation to convert alcohol content into kilocalories. However, some alcoholic drinks also contain a significant amount of carbohydrate (sugar, *e.g.* sweet, flavoured alcoholic beverages) and some contain saturated fats (*e.g.* cream liqueurs). Arguably, some alcoholic beverages might be allowed to declare just energy, but those containing a significant amount of other ingredients should be required to provide a full nutrition declaration (*i.e.* carbohydrate, fat, protein and salt).

The drinks industry has, therefore, been charged with developing a code of practice in this respect as well as how the information might be presented. Alcoholic drinks tend to have a long shelf life and, for spirits, liqueurs, *etc.*, a single bottle might be opened and consumed over a long period of time. One option might be for the information to be presented using links printed on the label or QR codes. However, this could make the information less accessible because many alcoholic drinks are sold in layered packaging, which can be lost or become illegible, or from behind public bars or other catering establishments.

1.8 Tolerances

Food is a difficult matrix and there are many reasons why the declared value of a nutrient might not correspond with values measured analytically. These include the breakdown of nutrients with time or through reactions with ambient oxygen (*e.g.* vitamin C). There might be variations in the sources of ingredients (*e.g.* tomatoes from the Netherlands *versus* Israel) or seasonal variations in natural components. Labels for a food are usually printed well before manufacture, but nonetheless food under FIC manufacturers are responsible for and must ensure information printed on the label is correct, particularly where there might be issues surrounding the over-consumption of a particular nutrient (*e.g.* energy) or a nutrition claim (*e.g.* vitamin C).

Given the potential for variation in their guidance, the EC indicated the following should be taken into account for enforcement: the nutrient in question, extent of the deviation, and whether the amount was an over- or under-estimation. If a product is claiming to be low fat or low sugar content then the tolerance would be greater, provided the product was below defined thresholds, than if the claim were "high in vitamin C" and the product was comparable with other foods without the claim.

Other aspects that would help in determining tolerance would be naturally high variation in nutrient content including seasonality, high degradation rate or analytical variability in some food matrices; low homogeneity leading to high variation of nutrient in sampling for analysis, compliance of the majority of samples from the lot with tolerance ranges, where such data are available; and the history of the company concerned, *i.e.* whether they are generally in compliance. To this end, the EC has published a table of tolerances for different nutrients (Table 1.1).[21]

Table 1.1 Tolerances for foods other than food supplements including measurement uncertainty.

	Tolerances for foods (includes uncertainty of measurement)	
Vitamins	+50%[a]	−35%
Minerals	+45%	−35%
Carbohydrate, sugars, protein, fibre	<10 g per 100 g: 10–40 g per 100 g: >40 g per 100 g:	±2 g ±20% ±8 g
Fat	<10 g per 100 g: 10–40 g per 100 g: >40 g per 100 g:	±1.5 g ±20% ±8 g
Saturates, monounsaturates, polyunsaturates	<4 g per 100 g: ≥4 g per 100 g:	±0.8 g ±20%
Sodium	<0.5 g per 100 g: ≥0.5 g per 100 g:	±0.15 g ±20%
Salt	<1.25 g per 100 g: ≥1.25 g per 100 g:	±0.375 g ±20%

[a]For vitamin C in liquids, higher upper tolerance values could be accepted.

For foods claiming to be high or low in a particular nutrient the differences on the side of the claim should reflect the claim. In other words, if the manufacturer claims to be high in a nutrient then there should only be measurement uncertainly should the nutrient fall below the claim. If the manufacturer is claiming to be low in a nutrient, then there should only be measurement uncertainty if the measurement is above the claim.

There are special conditions for the variability of nutrients in supplements because of their relatively long shelf life and the degradation over time.

Finally, there is a table giving the amount by which the amounts of the nutrients in food can be rounded to give more meaning to consumers.

References

1. M. Quigley, *Med. Law Rev.*, 2013, **21**(4), 588–621.
2. Regulation (EU) No 1169/2011 of the European Parliament and of the Council of 25 October 2011 on the provision of food information to consumers, http://data.europa.eu/eli/reg/2011/1169/oj [accessed August 2018].
3. European Commission Food Labelling Legislation, Nutrition and Health Claims, https://ec.europa.eu/food/safety/labelling_nutrition/claims_en [accessed August 2018].
4. European Commission Addition of vitamins and minerals, https://ec.europa.eu/food/safety/labelling_nutrition/vitamins_minerals_en [accessed August 2018].
5. Council Directive 90/496/EEC of 24 September 1990 on nutrition labelling for foodstuffs, http://data.europa.eu/eli/dir/1990/496/oj [accessed August 2018].
6. WHO, Global status report on noncommunicable diseases 2014, ISBN: 978 92 4 156485 4, http://www.who.int/nmh/publications/ncd-status-report-2014/en [accessed July 2018].
7. WHO, Global action plan for the prevention and control of NCDs 2013-2020, 2013, ISBN: 978 92 4 150623 6, http://www.who.int/nmh/publications/ncd-action-plan/en [accessed July 2018].
8. Dietary reference values and dietary guidelines, https://www.efsa.europa.eu/en/topics/topic/dietary-reference-values-and-dietary-guidelines [accessed August 2018].
9. CODEX General Standard for the Labelling of Prepackaged Foods, http://www.fao.org/docrep/005/Y2770E/y2770e02.htm [accessed August 2018].
10. Codex Committee on Food Labelling (CCFL), http://www.fao.org/fao-who-codexalimentarius/committees/committee/en/?committee=CCFL [accessed August 2018].
11. Regulation (EC) No 852/2004 of the European Parliament and of the Council of 29 April 2004 on the hygiene of foodstuffs, http://data.europa.eu/eli/reg/2004/852/oj [accessed August 2018].
12. Results of the Commission's consultations on 'trans fatty acids in food stuff in Europe', https://ec.europa.eu/food/sites/food/files/safety/docs/fs_labelling-nutrition_trans-fats-oswp_en.pdf [accessed August 2018].

13. UK Government Technical Guidelines for Traffic Light Labelling https://www.gov.uk/government/uploads/system/uploads/attachment_data/file/566251/FoP_Nutrition_labelling_UK_guidance.pdf [accessed July 2018].
14. G. Sacks, M. Rayner and B. Swinburn, *Health Promot. Int.*, 2009, **24**(4), 344–352.
15. Sainsbury's Healthy Labelling – Summary Charts December 2012, https://assets.publishing.service.gov.uk/government/uploads/system/uploads/attachment_data/file/213240/Sainsburys.pdf [accessed September 2018].
16. Guide to creating a front of pack (FoP) nutrition label for pre-packed products sold through retail outlets, https://assets.publishing.service.gov.uk/government/uploads/system/uploads/attachment_data/file/566251/FoP_Nutrition_labelling_UK_guidance.pdf [accessed September 2018].
17. Regulation (EC) No 1924/2006 of the European Parliament and of the council of 20 December 2006 on nutrition and health claims made on foods, https://eur-lex.europa.eu/legal-content/EN/ALL/?uri=CELEX%3A02006R1924-20121129 [accessed July 2018].
18. Reform of the common agricultural policy (CAP), https://eur-lex.europa.eu/legal-content/EN/TXT/?uri=LEGISSUM%3Al60002 [accessed August 2018].
19. Report from the Commission to the European Parliament and The Council regarding the mandatory labelling of the list of ingredients and the nutritional declaration of alcoholic beverages, https://ec.europa.eu/food/sites/food/files/safety/docs/fs_labelling-nutrition_legis_alcohol-report_en.pdf [accessed July 2018].
20. Regulation (EC) No 110/2008 of the European Parliament and of the Council of 15 January 2008 on the definition, description, presentation, labelling and the protection of geographical indications of spirit drinks and repealing Council Regulation (EEC) No 1576/89, http://data.europa.eu/eli/reg/2008/110(1)/oj [accessed August 2018].
21. Guidance Document for Competent Authorities for the Control of Compliance with EU Legislation with regard to the setting of tolerances for nutrient values declared on a label, https://ec.europa.eu/food/sites/food/files/safety/docs/labelling_nutrition-vitamins_minerals-guidance_tolerances_1212_en.pdf [accessed August 2018].

CHAPTER 2

Impact of Health Claims Regulations on the Food Industry and Future Innovation Potential

KATIA MERTEN-LENTZ

Keller and Heckman LLP, Avenue Louise 54, 1050 Brussels, Belgium
Email: mertenlentz@khlaw.com

2.1 Introduction

The food industry is one of the biggest European economic sectors in terms of added value. It is not surprising, therefore, that the competitiveness of this sector is a key objective for the European Union (EU) single market. However, the right to access safe and good quality food is also enshrined in EU legislation. Recent changes in consumption habits have augmented the role of quality and health benefits, leading to the introduction of claims that should primarily help consumers make a healthy, informed choices. Around a quarter of all food products in Europe carry nutrition and health claims. It could even be argued that health claims stimulate food innovation at the European level. The principles of nutrition and health claims use are described in Regulation (EC) No 1924/2006 on nutrition and health claims related to foods,[1] which stipulates requirements for scientific assessment and other elements that allow food business operators to make such claims on product packaging.

Food Chemistry, Function and Analysis No. 22
Health Claims and Food Labelling
Edited by Siân Astley

Published by the Royal Society of Chemistry, www.rsc.org

The Nutrition and Health Claims Regulation (NHCR) is part of the EU food legislation that has been harmonised. Regulation (EC) No 1924/2006 permits food business operators to benefit significantly from opportunities offered by the EU single market.[2] Health and nutrition claims are just one aspect of commercial communications made *via* food packaging, but their definitions depend on interpretation of the NHCR scope.[3] The use of nutrition and health claims is based on an authoritative statement, preceded by scientific assessment, and risk management decisions taken by the European Commission (EC).

2.2 From Food Safety Policy to Nutrition and Health Claim Regulation

The NHCR has been adopted as one of the actions undertaken by the EU to enforce food legislation. By harmonising legislation related to nutrition and health claims, at the European level, the EC aims to contribute to ensuring a high level of food safety.

2.2.1 General Food Law and Business Operators

In the 1990s, after several food scandals,[2] EU institutions decided to undertake action aimed at establishing the "highest standard of food safety".[3] Adoption of the 2002 General Food Law[4,5] was followed by a range of instruments with the principal objectives of ensuring food safety, regaining the trust of consumers, and providing a concrete legal framework for food business operators placing food products on the EU market.[5]

The General Food Law Regulation [Regulation (EC) No 178/2002] sets out several provisions that are the cornerstones to EU food law and create "a general framework of a horizontal nature to be followed when measures are taken"[6] as well as "the free movement of safe and wholesome food is an essential aspect of the internal market and contributes significantly to the health and well-being of citizens, and to their social and economic interests" (Recital 1). Thus, EU food legislation is a web of general principles and requirements, covering all stages of the food production chain as a continuum, *i.e.* from production up to sale or supply to the consumer.

2.2.2 Harmonisation of Nutrition and Health Claim Legislation

NHCR[7] was conceived in an attempt to strike a balance between the interests of food business operators and the protection of consumers. However, as a result, the food industry faced important burdens and legal uncertainties regarding the use of health claims. As a result, there was a need for new provisions on authorisation of claims.

2.2.2.1 Adoption of the Regulation

After many years of debate and discussion of around 700 amendments, the EU adopted a Regulation on nutrition and health claims on foods in 2006.[7] Rising living standards and increased rates of non-communicable diseases (*e.g.* obesity, type 2 diabetes) in past decades encouraged a functionalist approach towards nutrition, which was marked by a market expansion in products claiming health and nutrition benefits in order to appeal to consumers.[8]

Increasing numbers of food products with innovative features and new ingredients, and the use of health claims on such products, allowed some food business operators to increase sales and market shares. Gaining consumer attention is crucial, as it is "a bottle-neck for the use of health claims",[9] but it is important to ensure that innovations do not endanger consumer health and safety.[10] In addition, the NHCR aims to regulate the use of health and nutrition claims in the EU, simultaneously expanding opportunities for product innovation and promoting public health whilst protecting the interests of consumers with respect to misleading advertising.

2.2.2.2 Objectives: To Stimulate Markets and Protect Consumers

NHCR came into effect on 1 July 2007. Along with a high level of consumer protection, Regulation (EC) No 1924/2006 aims to facilitate consumer choices by making sure that claims are scientifically substantiated, based on a review conducted by the European Food Safety Authority (EFSA). NHCR also aims to improve free movement of goods in the EU by harmonising EU Member State (MS) legislation, and provide for fair competition in the food sector by increasing legal certainty for economic operators. Indeed, prior to Regulation (EC) No 1924/2006, MS had different approaches with regard to claims, ranging from an outright ban to the absence of any legislation. This lack of consistency resulted in fragmentation of the internal market. Moreover, some food business operators demonstrated questionable practices in the use of nutrition and health claims, meaning whilst some invested in research and development to substantiate nutrition and health claims, others used claims as a marketing tool without any scientific justification.[11]

These issues could only be resolved by addressing them at the EU level. Thus, to facilitate free movement of foods within the single market and avoid unfair competition and not jeopardise consumers' trust in scientifically justified claims, it was agreed that nutrition and health claims made on food packaging must be based on, and substantiated by, generally accepted scientific evidence. Also, health claims should only be authorised for use in the EU after a thorough scientific assessment (review) by the EFSA.

In its attempt to establish common rules for nutrition and health claims, which would improve legal certainty, contribute to free movement of food in the single market, and – most importantly – maintain health and safety standards for consumers, NHCR set ambitious objectives. However, not all of them have been realised. For example, it includes the introduction of EU-wide nutrient profiles to "avoid a situation where nutrition or health claims mask the overall nutritional status of a food product, which could mislead consumers when trying to make healthy choices in the context of a balanced diet", but their implementation is still pending, creating obstacles for the NHCR as a whole.

2.3 Scope of EU Nutrition and Health Claim Legislation

As stated in Article 1.2 of Regulation (EC) No 1924/2006, it applies to the "nutrition and health claims made in commercial communications, whether in the labelling, presentation or advertising of foods to be delivered as such to the final consumer, including foods which are placed on the market unpacked or supplied in bulk".

The European Court of Justice (ECJ) has played an important role in determining NHCR scope but, by interpreting it broadly, the result has been very restrictive on the use of health claims.

2.3.1 Claims

In Article 2, a claim is defined as "any message or representation, which is not mandatory under Community or national legislation, including pictorial, graphic or symbolic representation, in any form, which states, suggests or implies that a food has particular characteristics". It is clear from this definition that it covers not only explicit claims, but also visual claims, which are classified as implied health claims. Such a broad definition sanctions broad application of the NHCR in terms of the types of communication and products on which claims might be indicated, *i.e.* NHCR applies to nutrition and health claims made in commercial communications of food destined for the final consumer, which can be made in the form of labelling, presentation or advertising.

2.3.1.1 Commercial Aspects of Labelling

Nutrition and health claims are not mandatory under EU or national legislation and are linked closely to commercial objectives. Generally, nutrition and health claims refer to communications from the food business operator point of view, and have the aim of facilitating promotion of products and services, either directly or indirectly, thereby influencing buyers.[12] It should not be forgotten that such communications also include trademarks and brand names, which may be construed as nutrition or health claims (*e.g.* Activia Danone, 2010). Consequently, nutrition and health claims can be a strong marketing tool for food business operators.

The status of claims made in commercial communications and addressed exclusively to professionals has also been clarified by the ECJ (C-19/15). The Court stated, "if that communication is addressed not to the final consumer, but exclusively to health professionals, [it] falls within the scope of that Regulation".[13] It is important to note that claims made in non-commercial communications (*e.g.* dietary guidelines, advice issued by public health authorities and bodies, information in the press or in scientific publications, statements addressing the role of dietary patterns or of general categories of foods [*e.g.* fruits and vegetables]) are not within the scope of Regulation (EC) No 1924/2006. Objective information for health professionals about new scientific developments, using technical or scientific terminology, when the communication is of non-commercial nature is allowed, *i.e.* not within the scope of the Regulation.[14] However, due diligence is required, conducted on a case-by-case basis to determine whether claims made in business-to-business communications fall within the scope of NHCR. Whether claims take the form of labelling, presentation or advertising, the ECJ decided that "to be delivered as such to the final consumer" relates to the food and not the communication.[13]

2.3.1.2 Labelling of Food

NHCR covers all foods including natural mineral waters, water intended for human consumption and food supplements. It also includes foods placed on the market unpacked or supplied in bulk, foods intended for supply to restaurants, hospitals, schools, canteens and similar mass caterers (HoReCa). Importantly, medicinal claims are prohibited and their use on food supplements may render the product a medicine, under European Medicines Agency (EMA, www.ema.europa.eu) jurisdiction. However, the line between medical products and foods is narrow[15] and it is the responsibility of risk managers (*i.e.* EC) to decide whether a claim is medicinal and if individuals taking medications can be the target population for health claims made on foods.

NHCR also limits communication of claims for foods with an otherwise undesirable nutritional profile (*e.g.* high in fat, sugar or salt). Such foods may be consumed in moderation, as part of a healthy diet, but were these foods to include health or nutrition claims, many consumers might increase intakes inappropriately. The NHCR does not deem them "bad", but prevents them from being marketed as "good" with positive messages about health and nutritional benefits.

In this sense, setting nutrient profiles was supposed to establish healthy limits for different categories of foods.[16] Notwithstanding the fact that a number of nutrient profile models have been developed, nutrient profiles have yet to be established or implemented by the EC and MS.

2.4 From a Particular Characteristic to Nutrition and Health Claims

Particular characteristic means a health claim implies a relationship between a food category, food or constituent(s) and health whilst a nutritional claim links health with the energy and/or other nutrients that a food does or does not contain.

2.4.1 Nutrition Claims

More specifically, under Article 2, a nutrition claim means "any claim which states, suggests or implies that a food has particular beneficial nutritional properties due to the energy (calorific value) it provides; provides at a reduced or increased rate; or does not provide; and/or the nutrients or other substances it contains; contains in reduced or increased proportions; or does not contain;". Article 8 further specified that "[n]utrition claims shall only be permitted if they are listed in the Annex and are in conformity with the conditions set out in this Regulation". The EU list of permitted nutrition claims is outlined in the NHCR Annex, which also specifies conditions of use.[17]

2.4.2 Health Claims

Health claims are also defined in Article 2 as "any claim that states, suggests or implies that a relationship exists between a food category, a food or one of its constituents and health". There are three types of health claims: (1) function claims based on either accepted (Article 13.1) or (2) new scientific evidence (Article 13.5) and (3) Article 14.1 claims, which refer to the reduction of disease risk as well as children's development and health.

Function health claims (Article 13.1) refer to growth, development or function of the body, or to psychological, behavioural functions, or to slimming/weight control, hunger, satiety, reduction of available energy, *e.g.* "Vitamin D contributes to the normal function of the immune system". Such claims must be based on generally accepted scientific evidence and easily understood by an average consumer. Claims under Article 13.5 are based on newly developed scientific evidence, for which protection of proprietary data is also requested. For these health claims, authorisation is on a case-by-case basis, which reflects the investment by food operators in research and development costs.

Health claims under Article 14 are related to "reduction of disease risk" claims (Article 14.4) and claims referring to childhood development and health. In the case of risk reduction claims (*i.e.* reducing a risk factor in the development of a disease), product must also carry a statement indicating that the disease has multiple risk factors and altering just one may or may not have a beneficial effect (Article 14.2).

Thus, health claims have two essential components: a substance (food, food component or ingredient) and a disease/health-related condition. A positive relationship between the substance identified and improved health must be substantiated in a claim, *e.g.* "Vitamin D is needed for the normal growth and development of bone in children". The NHCR list describes authorised health claims, their conditions of use and applicable restrictions, if any.[17] It also states unauthorised health claims and the reasons for their exclusion.

2.4.3 Generic (Non-specific) Claims

Generic descriptors could be construed as claims, but they are not perceived as having beneficial effects. If anything, they indicate non-specific benefits of food for overall health and well-being and, consequently, are exempted from Article 1.4 but can only be included on packaging if accompanied by a health claim under Article 10, *e.g.* "heart-friendly" or "good for your stomach".

However, the ECJ considered claims sustaining beneficial physiological effect, contrary to other (similar) foods, which are presumed to result in sustained adverse effects on the digestive system and, consequently, on health, must be considered as a health claim.[18]

2.5 Scientific Substantiation for a Claim

Scientific substantiation is the main aspect considered in the use of nutrition and health claims (Recital 17). Substances subject to claims have to prove beneficial nutritional or physiological effects established by generally accepted scientific principles. Benefits or effects claimed must be linked to a significant quantity of substance or nutrient in the final product and the final product must be "bioavailable", *i.e.* ready for consumption. Beneficial effects may also be determined by the absence or reduction of a nutrient or substance (*e.g.* reduced cholesterol).

Health claims are only authorised for use after a scientific assessment (review) of the highest possible standard, which is carried out by EFSA and includes evaluation of risk(s) and benefit(s) of specified substances. However, authorisation of scientifically substantiated health claims may be withheld due to other factors.

2.6 Authoritative Statement of NHC

EFSA is responsible for ensuring the harmonised scientific assessment (review) of claims and, upon request, applicants can access applications to check on their status (Recital 23). A positive authoritative statement depends wholly on the quality and strength of the scientific evidence. However, the EC and the MS can intervene to assess risk management of the claim and deliver a negative decision, regardless of a positive opinion from EFSA on the scientific validity of the claim.

2.6.1 Application

An application shall include information on the applicant as well as on the nutrient (or another substance) or the food (or the category of food) about which the health claim is to be made and its specific characteristics, as well as any studies available demonstrating the health claim complies with criteria provided in Article 15 of the NHCR. Owing to the scientific and technical complexities of health claims, EFSA's Panel on Dietetic Products, Nutrition & Allergies has developed scientific criteria for substantiation of health claims. The latest guidance was published in January 2016[19] and outlined the evaluation process with the intention to clarify assessment criteria and resolve contradictions in some of the basic scientific principles.

In order to demonstrate the relationship between consumption of the food/constituent and maintenance of the function (claimed effect), high quality information and studies supporting the claim must be submitted. Applications must demonstrate that the nutrient is essential and be characterised fully, and the claimed effect formulated based on human studies identifying the outcome(s) and beneficial physiological effect(s). A rationale (biological mechanism) should be developed to substantiate plausibility of the claim with supporting evidence from the studies. The wording and conditions of use for the claim must be defined and should reflect the scientific evidence, but consumer understanding is not a requirement for the scientific assessment.[20] Proving scientifically health benefits of a certain food, food component or ingredient is complex[21] and resource intensive for food business operators, and shown to be a barrier for many manufacturers, particularly small- and medium-sized enterprises (SMEs), which make up the majority of food manufacturers in Europe.

2.6.2 From Risk Assessment to Risk Management

Application for authorisation must be submitted to the national competent authority of a MS, which acknowledges receipt and informs EFSA without delay. The EFSA working group will discuss and evaluate a draft opinion on the claim and will submit it to the Panel for discussion and adoption. Then, it publishes a scientific opinion forming the basis of the final decision published by the EC. Two months after receiving the opinion from EFSA, the EC should submit a draft decision to the Standing Committee on Plants, Animals, Food and Feed (Article 17.1). That draft decision should consider the opinion of EFSA "and other legitimate factors", such as safety (*e.g.* modify conditions/restrictions of use) and consumer understanding (*e.g.* modify the wording). The draft decision must be adopted in accordance with the Regulatory Procedure with Scrutiny. However, in practice because other factors must be considered, there are often discrepancies between

EFSA positive opinions and EC final decisions, *i.e.* health claims might be rejected despite a positive opinion from EFSA, which ensures a high level of consumer protection.[22]

2.6.3 Nutrient Profiles Issue

NHCR stipulated that the EC should establish nutrient profiles by January 2009 to ensure the good nutritional status of foods and support consumers in making healthier choices. Whilst being based on the scientific opinion of EFSA, nutrient profiles were intended to allow innovation, and embrace variations in dietary habits and traditions. Thus, nutrient profiles are not a scientific concept, but a regulatory tool based on scientific and non-scientific elements.[23]

Nutrient profiles should be established for foods and/or certain categories of food, considering amounts of priority nutrients (energy, fats, saturated fats, protein and sugars) and other substances (*e.g.* vitamins and minerals), role and importance of the food, contribution to the eating habits of consumers, overall nutritional composition, and the presence of nutrients proven scientifically to influence health. To assess whether profiles should be set for foods in general and/or for categories of foods, EFSA should issue an opinion and consult with stakeholders. However, the issue of setting nutrient profiles has raised considerable political interest and generated intense debate. Despite the fact that EFSA adopted an opinion (31 January 2008), the EC draft proposal has never reached the Standing Committee. The European Parliament (Legal Affairs Committee) proposed to eliminate requirements related to nutrient profiles,[24] but the EC evaluated this Regulation as part of the Better Regulation Communication (19 May 2015). In the roadmap (October 2015)[11] published, describing the effectiveness, coherence and added value of setting nutrient profiles, the EC questioned the impact of not setting nutrient profiles at the EU level, regarding realisation of NHCR objectives, but also costs and benefits for the food sector, as well as identifying alternatives to achieve similar objectives. The outcome of this evaluation was expected in the first quarter of 2018[25] and, in the meantime, 262 health claims and nutrition claims have been listed in the NHCR Annex without nutrient profiles being established, despite several national and private initiatives on nutrient profiles schemes.

The continued absence of an EU regulatory framework in respect of nutrient profiles may impede innovation by the food industry, whereas adoption of nutrient profiles, with achievable criteria, would encourage product innovation, spurring food business operators to reformulate products or develop new ones to fulfil the requirements of nutrient profiles. Indeed, nutrient profiles could level the playing field for food business operators, allowing companies to compete using nutritional qualities rather than price or taste. Also, healthier products, achieved through voluntary reformulation, would bring longer-term predictability to the market, which is essential for investment in health-related innovation across the entire food industry.[26]

In addition, supporting consumers to make more informed choices about their food consumption has been undermined by the absence of nutrient profiles, as it allows nutrition or health claims for some foods containing large amounts of saturated fats, sugars or salt and trans fats. This leads to the perception that such foods are healthier whereas, in fact, their consumption is contrary to dietary recommendations. Thus, scientific assessment of nutrition and health claims is limited by the continued absence of EU-wide nutrient profiles, creating legal uncertainty and difficulties for food business operators, regardless of whether they want to include claims of this sort or not.

2.7 Permitted Claims

2.7.1 Permitted Nutrition Claims

Nutrition claims are only permitted if they are listed in the NHCR Annex. To date, 24 nutrition claims have been permitted, *e.g.* "with no added sugar", "high fibre", "contains (name of the nutrient or other substance)". Each of the permitted claims has specified conditions, and some nutrition claims are prohibited, such as "90% fat free", "starch reduced" or "low carb".

2.7.2 Permitted Health Claims

The list of authorised claims was adopted in 2012. Since then, food business operators have gained experience in the practical aspects of health claims. However, legal uncertainties remain, driving additional costs and retraining the sector dynamic. This is illustrated by the fact that EFSA and EC assessment criteria have been contested before the ECJ, but the Courts refused to question the NHCR, despite acknowledged gaps[27] in the Regulation. A favourable scientific opinion from EFSA does not guarantee authorisation, and the process of proving health benefits is complex and resource intensive for claims based on newly developed scientific data (Article 13.5) or reduction of disease risk (Article 14.1). Thus, food innovation is impeded and food business operators with longer-term health and innovation strategies have an advantage over SMEs when building evidence to support a claim application.[21]

2.7.3 Suspension Claims

More than 2000 claims have not been evaluated by EFSA or are still waiting to be considered by the EC on the so-called "list of health claims on hold". According to the EC, these claims, particularly those concerning the effects of plant and herbal substances (botanical substances) remain on hold and can continue to be used, in accordance with the transitional scheme in Article 28 of the NHCR. The biggest problem for these compounds is that, currently, there is no legally binding, EU-wide definition of botanicals. They may include plants and plant preparations[28] and/or herbs, plants, fungi, algae and extracts thereof with nutritional or physiological effect(s).[29] They may be considered to be traditional herbal medicine products or foods, but this conflict of legislation creates potential discrimination in the market and confusion amongst consumers. If botanicals are treated as foods, clinical trials must be submitted to EFSA and traditional use will not be considered, which explains they are systematically rejected benefiting neither the food industry nor the consumer. Regardless, botanicals remain on hold at the EC's discretion, in accordance with Regulation No 1924/2006 and in EU law.[30]

NHCR will have a significant effect on harmonisation of other ingredients in food supplements, which constitute an important aspect of food innovation, most notably in the light of adoption of the novel food regulation.[31–33] Regulation (EU) No 2015/2283 lays down novel food categories and considers new food technologies as well as a specific procedure for traditional ingredients used in third countries (*e.g.* traditional plant products). From a regulatory point of view, The European Commission's REgulatory FItness and Performance Programme (REFIT)[34] should examine whether current rules concerning health claims for plant compounds used in foods are adequate. This would allow the evaluation of the botanicals to be (1) added to the current rules, (2) withdrawn from the scope of NHCR or (3) development of specific legislation.

2.8 Conclusion

The use of nutrition and health claims is tied intrinsically to freedom to conduct a business. The ECJ found that restrictions imposed by the NHCR, regarding the use of health claims, corresponded to objectives of general interest pursued by the EU and do not constitute a disproportionate and intolerable interference, even if they are based on uncertain criteria, *e.g.* "other factors".[35] However, Regulation (EC) No 1924/2006 (NCHR) appears to be impaired, considering the number of rejected and pending applications for health claims, the issues around nutrient profiles, and suspension of botanicals from the authorisation process. Considering recent legal evolutions regarding food innovations [*e.g.* novel food regulation, Regulation (EU) No 2015/2283] and rising numbers of new models for food consumption (*e.g.* personalised nutrition), it is important for food business operators to have a concrete and certain legal framework to stimulate innovation and ensure consumer protection and food safety. In its roadmap, the EC proposed to examine whether nutrient profiles, provided for in the Regulation but not yet adopted, are warranted and adequate to ensure the objectives of the Regulation, and examine whether current rules concerning health claims on plant compounds used in foods are adequate. However, for the moment, the outcomes of REFIT and the legal uncertainties of NCHR remain, potentially stifling innovation and not protecting the interests of consumers as intended.

References

1. Regulation (EC) No 1924/2006 of the European Parliament and of the Council of 20 December 2006 on nutrition and health claims made on foods, http://data.europa.eu/eli/reg/2006/1924/oj [accessed August 2018].
2. A. de Boer, M. Urlings, E. Vos and A. Bast, *European Food and Feed Law Review*, 2015, **10**(5), 334–344.
3. European Commission, White Paper on Food Safety, https://ec.europa.eu/food/sites/food/files/safety/docs/animal-feed-pub06_en.pdf [accessed July 2018].
4. European Commission General Food Law, https://ec.europa.eu/food/safety/general_food_law_en [accessed August 2018].
5. Regulation (EC) No 178/2002 of the European Parliament and of the Council of 28 January 2002 laying down the general principles and requirements of food law, establishing the European Food Safety Authority and laying down procedures in matters of food safety, http://data.europa.eu/eli/reg/2002/178/oj [accessed August 2018].
6. Judgment of the Court (Fourth Chamber) of 19 January 2017, Queisser Pharma GmbH & Co. KG v Bundesrepublik Deutschland, ECLI:EU:C:2017:26 (paragraph 49), http://curia.europa.eu/juris/document/document.jsf;jsessionid=9ea7d2dc30d83a6b9f7973a94ef398cb9a73bd717aaa.e34KaxiLc3qMb40Rch0SaxyPahr0?text=&docid=186968&pageIndex=0&doclang=en&mode=lst&dir=&occ=first&part=1&cid=507565 [accessed August 2018].
7. Regulation (EC) No 1924/2006 of the European Parliament and of the council of 20 December 2006 on nutrition and health claims made on foods, http://data.europa.eu/eli/reg/2006/1924/2012-11-29 [accessed August 2018].
8. M. Friant-Perrot, *Revue de Droit Rural*, 2013, 19.
9. Final Report Summary – CLYMBOL (Role of health-related claims and symbols in consumer behaviour), https://cordis.europa.eu/result/rcn/195496_en.html [accessed July 2018].
10. A. Tarabella and B. Burchi, *Procedia – Soc. Behav. Sci.*, 2012, **46**, 2173–2177.
11. Evaluation of (a) Regulation (EC) No 1924/2006 on nutrition and health claims made on food with regard to nutrient profiles and health claims made on plants and their preparations and of (b) the general regulatory framework for their use in foods, http://ec.europa.eu/smart-regulation/roadmaps/docs/2015_sante_595_evaluation_health_claims_en.pdf [accessed August 2018].
12. Opinion of advocate general Saugmadsgaard ØE, 18 February 2016 on case C-19/15, Verband Sozialer Wettbewerb e.V. v Innova Vital GmbH, ECLI:EU:C:2016:563, paragraph 39, http://curia.europa.eu/juris/document/document.jsf?text=&docid=181681&doclang=EN [accessed August 2018].
13. Judgment of the Court (Third Chamber) of 14 July 2016, Verband Sozialer Wettbewerb eV v Innova Vital GmbH, ECLI:EU:C:2016:563,

http://curia.europa.eu/juris/liste.jsf?pro=&lgrec=en&nat=or&oqp=&dates=&lg=&language=en&critereEcli=ECLI%253AEU%253AC%253A2016%253A563&jur=C%2CT%2CF&cit=none%252CC%252CCJ%252CR%252C2008E%252C%252C%252C%252C%252C%252C%252C%252C%252C%252Ctrue%252Cfalse%252Cfalse&td=%3BALL&pcs=Oor&avg=&page=1&mat=or&jge=&for=&cid=515976 [accessed August 2018].
14. S. Roset, *Revue Europe*, 2016, **10**, 342.
15. Judgment of the Court (Third Chamber) of 23 November 2016, Nelsons GmbH v Ayonnax Nutripharm GmbH and Bachblütentreff Ltd, ECLI:EU:C:2016:888, http://curia.europa.eu/juris/liste.jsf?language=en&num=C-177/15 [accessed August 2018].
16. The setting of nutrient profiles for foods bearing nutrition and health claims pursuant to Article 4 of the Regulation (EC) No 1924/2006 – Scientific Opinion of the Panel on Dietetic Products, Nutrition and Allergies, https://www.efsa.europa.eu/en/efsajournal/pub/644 [accessed August 2018].
17. European Commission Nutrition Claims, https://ec.europa.eu/food/safety/labelling_nutrition/claims/nutrition_claims_en [accessed August 2018].
18. Judgment of the Court (Third Chamber), 6 September 2012, Deutsches Weintor eG v Land Rheinland-Pfalz, ECLI:EU:C:2012:526, http://curia.europa.eu/juris/liste.jsf?pro =&lgrec=en&nat=or&oqp=&dates=&lg=&language=en&jur=C%2CT%2CF&cit=none%252CC%252CCJ%252CR%252C2008E%252C%252C%252C%252C%252C%252C%252C%252C%252C%252Ctrue%252Cfalse%252Cfalse&num=C-544%252F10&td=%3BALL&pcs=Oor&avg=&page=1&mat=or&jge=&for=&cid=517061 [accessed August 2018].
19. EFSA General scientific guidance for stakeholders on health claim applications, https://www.efsa.europa.eu/en/efsajournal/pub/4367 [accessed August 2018].
20. EFSA Scientific and technical guidance for the preparation and presentation of a health claim application (Revision 2), https://www.efsa.europa.eu/en/efsajournal/pub/4680 [accessed August 2018].
21. D. Ferguson, Researchers show how to navigate complex EU health claim regulations for food and drinks, https://www.surrey.ac.uk/mediacentre/press/2018/researchers-show-how-navigate-complex-eu-health-claim-regulations-food-and [accessed August 2018].
22. Judgment of the Court (Eighth Chamber) of 8 June 2017, Dextro Energy GmbH & Co. KG v European Commission, ECLI:EU:C:2017:437, Paragraph 80, http://curia.europa.eu/juris/liste.jsf?language=en&jur=C,T,F&num=C-296/16%20P&td=ALL [accessed August 2018].
23. Answer given by Mr Andriukaitis on behalf of the Commission 17 July 2015 to parliamentary question E-008301/2015, http://www.europarl.europa.eu/sides/getAllAnswers.do?reference=E-2015-008301&language=EN [accessed August 2018].

24. A. Harrison-Dunn, EU Parliament could ask Commission to ditch nutrient profiles for health claims, https://www.nutraingredients.com/Article/2015/06/09/EU-Parliament-could-ask-Commission-to-ditch-nutrient-profiles-for-health-claims [accessed August 2018].
25. (a) Answer given by Mr Andriukaitis on behalf of the Commission the 14th of February 2017 to Parliamentary question E-009744/2016, http://www.europarl.europa.eu/sides/getAllAnswers.do?reference=E-2016-009744&language=EN [accessed August 2018]; (b) Answer given by Mr Andriukaitis on behalf of the Commission 26 May 2015 E-005223/2015, http://www.europarl.europa.eu/sides/getAllAnswers.do?reference=E-2015-005223&language=EN [accessed August 2018].
26. Impact of EU Regulation on Innovation: Repository of Industry Cases, Business Europe Report, December 2016, https://www.businesseurope.eu/sites/buseur/files/media/reports_and_studies/2016-12-02_impact_of_eu_regulation_on_innovation_-_repository_of_industry_cases.pdf [accessed August 2018].
27. Judgment of the General Court (Eighth Chamber) of 12 June 2015, The Health Food Manufacturers' Association and Others v European Commission, http://curia.europa.eu/juris/liste.jsf?language=en&num=T-296/12 [accessed August 2018].
28. Decree No. 2006-352 20 March 2006 Relating to Food Supplements, https://www.global-regulation.com/translation/france/1654421/decree-no.-2006-352-20-march-2006-relating-to-food-supplements.html [accessed August 2018].
29. The Use of Substances with Nutritional or Physiological Effect Other Than Vitamins and Minerals in Food Supplements, Study Undertaken for DG SANCO, European Commission, 2007, https://ec.europa.eu/food/sites/food/files/safety/docs/labelling_nutrition-supplements-2007_a540169_study_other_substances_en.pdf [accessed August 2018].
30. Order of the Court (Eighth Chamber) of 25 October 2016, VSM Geneesmiddelen BV v European Commission, http://curia.europa.eu/juris/liste.jsf?num=C-637/15&language=en [accessed August 2018].
31. Regulation (EU) 2015/2283 of the European Parliament and of the Council of 25 November 2015 on novel foods, amending Regulation (EU) No 1169/2011 of the European Parliament and of the Council and repealing Regulation (EC) No 258/97 of the European Parliament and of the Council and Commission Regulation (EC) No 1852/2001, http://data.europa.eu/eli/reg/2015/2283/oj [accessed August 2018].
32. European Commission, Novel Food, https://ec.europa.eu/food/safety/novel_food_en [accessed August 2018].
33. 1169/2011 of the European Parliament and of the Council and repealingRegulation (EC) No 258/97 of the European Parliament and of the Council and Commission Regulation (EC) No 1852/2001, http://data.europa.eu/eli/reg/2015/2283/oj [accessed August 2018].

34. Refit – making EU law simpler and less costly, https://ec.europa.eu/info/law/law-making-process/evaluating-and-improving-existing-laws/refit-making-eu-law-simpler-and-less-costly_en [accessed August 2018].
35. (A) Judgment of the Court (Third Chamber), 6 September 2012, ECLI:EU:C:2012:526, Paragraph 54, (B) Judgment of the Court (Sixth Chamber) of 15 April 1997, The Irish Farmers Association and others v Minister for Agriculture, Food and Forestry, Ireland and Attorney General, ECLI:EU:C:1997:187, Paragraph 27 (C) Joined Cases C-20/00 and C-64/00 Booker Aquaculture and Hydro Seafood [2003] ECR I-7411, paragraph 68.

CHAPTER 3

Supporting the European Food Industry to Meet Nutrient Labelling Regulations

MARIE MACHÁČKOVÁ,*[a] MARK ROE,[b] MARIJA RANIC[b] AND PAUL FINGLAS[b,c]

[a] Institute of Agricultural Economics and Information, Manesova 75, Prague 2 120 00, Czech Republic; [b] EuroFIR AISBL, 40 rue Washington, 1050 Brussels, Belgium; [c] Quadram Institute Bioscience, Norwich Research Park, NR4 7UA, UK
*Email: machackova.marie@uzei.cz

3.1 Introduction

In the European Union (EU), mandatory nutrition labelling has been regulated by Regulation (EU) No 1169/2011 on the provision of food information to consumers (hereafter, FIC)[1] since 13 December 2016. Prior to FIC, EU nutrient labelling was regulated by the Council Directive No. 90/496/EEC on nutrition labelling for foodstuffs and its amendments.[2] However, use of a nutrition declaration was voluntary, unless a nutrition-related claim was also present.

In the EU, FIC has been a milestone in nutrition labelling with mandatory rather than voluntary nutrition labelling being the most important impact.[3] For the first time, high-level unification and harmonisation of European rules for mandatory and voluntary provisions of food and nutrient labelling have been defined. The key element of FIC was that labelling controls

Food Chemistry, Function and Analysis No. 22
Health Claims and Food Labelling
Edited by Siân Astley

Published by the Royal Society of Chemistry, www.rsc.org

changed from a Directive to a Regulation, which meant interpretation of FIC rules into national legislation, was not required. FIC became applicable at the same time and in the format in all 28 EU Member States (EU MS),[4] which is crucial for free movement of goods across the EU Single Market.

Application of FIC was in two stages: during the transition period (13 December 2014 to 12 December 2016), voluntary nutrient labelling was allowed, provided it was in compliance with FIC, but from 13 December 2016, mandatory nutrient labelling came into force across the EU. The 28 EU MS have extended regions with mandatory nutrition labelling, following a global trend towards mandatory nutrition labelling.[5,6] The Food and Agriculture Organization (FAO) and World Health Organization (WHO) *Codex Alimentarius* guidelines on nutrition labelling reflect this, recommending that nutrition labelling should be mandatory, even in the absence of nutrition or health claims.[7] The European Food Information Council's 2018 edition of Global Update on Nutrition Labelling reports the following countries have made nutrition labelling mandatory: "United States, Canada, Mexico, Argentina, Brazil, Chile, Colombia, Ecuador, Paraguay, Uruguay, European Union (EU) Member States, Russia, Israel, Gulf Cooperation Council members, Nigeria, India, Hong Kong, China, Japan, South Korea, Malaysia, Taiwan, Thailand, Philippines, Indonesia, Vietnam, Australia and New Zealand".[5]

Differences between label formats and permitted claims among countries do exist and cause confusion among consumers as well as contributing to trade barriers for food manufacturers and distributors. The introduction of EU harmonised nutrient labelling is a good example of how regional disparities in nutrient labelling can be removed. The most recent initiative in regional harmonisation has been launched by the Association of Southeast Asian Nations (ASEAN), starting with discussions and analysis of possible approaches to addressing the challenges associated with provision of information for consumers.[8,9]

There remains, however, a predicament, namely that "mandatory provision of nutrition information on packaging should assist nutrition actions as part of public health policies, which could involve the provision of scientific recommendations for nutrition education for the public and support informed food choices" (ref. 1, Recital 34). To deliver this and comply with the legislation, food business operators (FBOs) must undertake measures to implement mandatory and voluntary particulars of nutrient labelling, as they are legally responsible for ensuring the information provided for consumers is correct. However, it is essential that FBOs understand the legislation and know how it applies to their business, whether they access it directly or through other organisations, *e.g.* trade associations, advisors, consultancies or other channels.

While larger FBOs may employ experienced staff with specific responsibilities that include labelling compliance, the majority of European FBOs are small and medium-sized enterprises (SMEs) and face considerable challenges in complying with labelling requirements. These FBOs have neither the personnel nor the knowledge and experience necessary (*e.g.* nutritional content based on laboratory analysis or recipe calculation from national food composition tables). Although legislation and guidelines are available online, even when simplified they can be confusing for users who do not have a technical background. Thus, information and services to support EU FBOs in implementing nutrition labelling are essential.

This chapter considers the information provided, and actions undertaken, by competent authorities, trade associations and others across the EU or within EU MS to inform FBOs not only of their responsibilities but also what they must do to deliver accurate nutrient labelling that is also compliant with FIC. Aspects considered include access to legislation documents and the availability of guidelines and training around nutrition labelling practices. The significance of consumer education, as the target group for information provided on labels, is also considered briefly.

3.2 Access to Sources of Food Law

Access to authoritative sources of food law is a prerequisite for application in practice. In this section, sources of information about food law at EU and national levels are discussed briefly.

3.2.1 European Union

3.2.1.1 The EUR-Lex Portal

The EUR-Lex portal[10] provides access to all EU law and other public documents in the 24 official languages of the EU. The portal also allows users to follow procedures culminating in the adoption of legal acts. An official compendium of all EU legal acts, the Official Journal (OJ), is the main source of EUR-Lex content. Electronic editions of the OJ (e-OJ) published after 1 July 2013 have legal force.

FIC pages[1] include the text of the Regulation, published electronically from the OJ, with information about procedures, dates of effect, and access to related documents (including amendments, consolidated versions, documents based on FIC, *e.g.* implementing acts and legislative procedures). The text also includes exceptions that relate mainly to minimally processed foods and those with little nutritional value (*e.g.* unprocessed foods consisting of a single ingredient, herbs and spices, sweeteners, tea and herbal and fruit infusions, waters, flavourings). Food supplied directly in small quantities to the final consumer by a manufacturer or to local retail establishments supplying the final consumer is also exempt. Interpretation of "small quantities" varies between EU MS but many, for example the UK, use the EU definition[11] of a micro-business, *i.e.* fewer than 10 employees and a turnover/balance sheet total of less than €2 m. Definitions of "local" also vary between EU MS, according to local definitions (*e.g.* administrative regions, provinces, districts, counties).

3.2.1.2 Nutrition Labelling Issues on the Official European Commission Portal

The framework for EU food and nutrition labelling issues is presented under the Food Section of the Official European Commission website,[12] where the following are also considered:

- Food labelling legislation
- Nutrition and health claims
- Addition of vitamins and minerals
- Natural mineral waters
- Food for specific groups

FIC is characterised in detail on the food labelling legislation pages[13] including key changes, information about the adoption of FIC, background papers and impact assessment reports. These pages provide materials about application of FIC (Question & Answers – see Section 4.3.1.1)[14,15] and guidance documents on nutrition labelling for competent authorities including tolerances for control of compliance with nutrient values (see Section 4.3.2).[16]

3.2.2 EU MS

FIC is applicable in all 28 EU MS. In addition to mandatory and voluntary nutrition labelling described in FIC, EU MS may adopt national measures (*e.g.* non-prepacked foods, mandatory additional information [*e.g.* vitamins and minerals], exceptions).[1] Conditions for the introduction of national measures are specified in FIC (ref. 1, Chapter VI). EU MS may not adopt nor maintain national measures unless authorised within FIC or elsewhere (ref. 1, Article 38(1)).

The implications of EU food laws and statutes instituted within the EU in response to FIC have been reviewed elsewhere by a group of national law experts.[17,18] The key finding was that, despite the large measure of harmonisation achieved, FIC does not cover all aspects of food information law. Some areas were left deliberately to national legislators in the belief that national regulators are better placed to judge what information consumers need to receive, *i.e.* "depending on local practical conditions and circumstances" (ref. 1, Recital 48), specifically with regard to non-prepacked foodstuffs. The major benefit of this approach is flexibility, recognising the potential impact on SMEs and enabling national regulators to manage this locally.

3.2.2.1 Collection of EU Member State Laws

In addition to the EUR-Lex portal, there are national online resources, summarising EU MS requirements, which can be accessed *via* N-Lex, an application within EUR-Lex.[19] N-Lex supports searching through a standardised form (in all 24 official EU languages). It is also important to consider practical application of the regulation and enforcement practice of EU MS, *i.e.* focusing often limited resources for compliance monitoring on products and FBOs that are considered to be higher risk for public health.

3.2.2.2 *Nutrition Labelling Issues Covered by National Authorities of the EU MS*

The various national authorities provide information about nutrient labelling (Table 3.1). In the majority of EU MS, this information is presented in a framework comprised of:

- General introduction to FIC
- Guidelines describing individual particulars of FIC
- Links to:
  - Text in the EUR-Lex portal[1]
  - Questions and Answers document on FIC, prepared by the European Commission (EC) (see Section 4.3.1.1)[14]
  - Tolerances document (see Section 4.3.2)[16]
  - Explanation of other EU regulation provisions related to nutrition labelling
- A list of national food laws and/or related EU regulations
- National interpretations related to FIC, *e.g.* definition of "local" or "small quantities"
- Contact for submitting queries

3.3 Guidelines

Application of FIC for nutrition labelling is not easy. FBOs must consider the scope of requirements, especially if they decide to apply supplementary information. EU MS authorities, trade associations and other bodies have published guidelines intended to help FBOs comply with the Regulation. These focus predominantly on the layout of nutrition labelling, specifically position and content of the nutrition declaration. There are also guidelines on obtaining nutrient values, which focus mainly on calculation of nutrient values. Examples of the different guidelines, developed by governmental authorities and the industry, are listed in Table 3.2.

3.3.1 Guidelines: Formats

3.3.1.1 *Question and Answers*

The EC has used a question and answer (Q&A) format for clarification of some provisions on the application of FIC. The document was elaborated by a Working Group, comprised of experts from EU MS, and the Q&A were

Table 3.1 National authorities in MS providing guidance on application of FIC, as related to nutrition labelling.

Country	Authority	Link (all accessed November 2018)
Austria	Federal Ministry of Health and Women's Affairs	https://www.bmgf.gv.at/home/Gesundheit/VerbraucherInnengesundheit/Lebensmittel/EU_Vorschriften_zur_Lebensmittelkennzeichnung#f1
Belgium	Belgian Federal Public Service Health, Food Chain Safety and Environment	https://www.health.belgium.be/en/node/23253
Bulgaria	Bulgarian Food Safety Agency	http://bfsa.bg/bg/Page/118/index/118
Croatia	Ministry of Agriculture	https://poljoprivreda.gov.hr/istaknute-teme/hrana-111/oznacavanje-hrane/informiranje-potrosaca-o-hrani/227
Cyprus	Ministry of Health – Food Safety Council	https://www.moh.gov.cy/Moh/fsc/fsc.nsf/All/FE4CB348D7F9D1ADC2257DB6004798A4?OpenDocument&highlight=1169/2011
Czech Republic	Ministry of Agriculture – Food Safety Authority or Czech Agriculture and Food Inspection Authority	http://www.bezpecnostpotravin.cz/kategorie/oznacovani-potravin.aspx or http://www.szpi.gov.cz/clanek/vyzivova-nutricni-hodnota-kdy-a-jak-ma-byt-oznacena-na-obalu.aspx?q=JmNobnVtPTImaGw9bnV0cmlkYXRhYmF6ZQ%3d%3d
Denmark	Danish Veterinary and Food Administration (DVFA)	https://www.foedevarestyrelsen.dk/Selvbetjening/Vejledninger/naeringsdeklarationsvejledning/Sider/default.aspx
Estonia	Ministry of Rural Affairs	https://www.agri.ee/et/uudised-pressiinfo/toidumargistusest-tarbijale
Finland	Ministry of Agriculture and Forestry	https://mmm.fi/pakkausmerkinnat-ravitsemus-ja-terveysvaitteet
France	General Directorate for Competition Policy, Consumer Affairs and Fraud Control	https://www.economie.gouv.fr/dgccrf/etiquetage-des-denrees-alimentaires-nouvelles-regles-europeennes
Germany	Federal Ministry of Food and Agriculture	https://www.bmel.de/DE/Ernaehrung/Kennzeichnung/kennzeichnung_node.html
Greece	Hellenic Food Authority	http://www.efet.gr/portal/page/portal/efetnew/legislations/sociable_legislations/food_notification
Hungary	National Food Safety Authority	http://portal.nebih.gov.hu/-/2016-december-13-tol-kotelezo-lesz-az-elelmiszereken-a-tapertekjeloles
Ireland	Food Safety Authority of Ireland	https://www.fsai.ie/legislation/food_legislation/food_information_fic/nutrition_labelling.html
Italy	Ministry of Economic Development	http://www.sviluppoeconomico.gov.it/index.php/it/impresa/competitivita-e-nuove-imprese/industria-alimentare#et
Latvia	Ministry of Agriculture Republic of Latvia	https://www.zm.gov.lv/partika/statiskas-lapas/partikas-markesana?id=6195#jump

Table 3.1 (*Continued*)

Country	Authority	Link (all accessed November 2018)
Lithuania	Ministry of Health of The Republic of Lithuania	https://sam.lrv.lt/lt/veiklos-sritys/visuomenes-sveikatos-prieziura/mityba-ir-fizinis-aktyvumas-2/informacijos-apie-maista-teikimas-maisto-produktu-zenklinimas
Luxembourg	Food Safety Agency	https://securite-alimentaire.public.lu/fr/professionnel/Denrees-alimentaires/Etiquetage.html
Malta	Ministry for Health	https://deputyprimeminister.gov.mt/en/environmental/Legislation/Pages/Food-Legislation/EU/Food-Information-to-Consumers.aspx
Netherlands	The Netherlands Food and Consumer Product Safety Authority	https://www.nvwa.nl/onderwerpen/etikettering-van-levensmiddelen
Poland	Agricultural and Food Quality Inspection	https://ijhars.gov.pl/przewodnik-do-rozporzadzenia-11692011.html
Portugal	The Ministry of Agriculture, Forestry and Rural Development	http://www.dgv.min-agricultura.pt/xeov21/attachfileu.jsp?look_parentBoui=6889501&att_display=n&att_download=y
Romania	Romanian Consumer Protection Authority	http://www.anpc.gov.ro/anpcftp/interes_public/ghid_150226.pdf
Slovakia	Ministry of Agriculture and the State Veterinary and Food Administration of the Slovak Republic	http://www.mpsr.sk/index.php?start&navID=1&id=5189 or https://www.svps.sk/dokumenty/potraviny/Otazky_odpovede_1.pdf
Slovenia	Ministry of Agriculture, Forestry and Food	http://www.uvhvvr.gov.si/si/delovna_podrocja/zivila/oznacevanje_zivil/
Spain	Spanish Agency for Consumer Affairs, Food Safety and Nutrition	http://www.aecosan.msssi.gob.es/AECOSAN/web/seguridad_alimentaria/subdetalle/norma_general_etiquetado.htm and http://eletiquetadocuentamucho.aecosan.msssi.gob.es/
Sweden	National Food Agency	https://www.livsmedelsverket.se/om-oss/lagstiftning1/gallande-lagstiftning/eu-forordning-11692011/
UK	Department of Health and Social Care	https://www.gov.uk/government/publications/technical-guidance-on-nutrition-labelling

Table 3.2 Examples of different guidelines developed by governmental authorities and industry.

Country	Government	Industry
European Commission	A working group launched by SANCO prepared: Questions & Answers (COM Q&A) on application of the FIC Regulation (January 2013) On 31 May 2018, the Commission adopted a Commission Notice to COM Q&A covering questions raised after entry into application of FIC, https://ec.europa.eu/food/safety/labelling_nutrition/labelling_legislation_en A Guidance Document (TOLERANCES) for competent authorities for Control of Compliance with EU Legislation [...] with regard to the setting of tolerances for nutrient values declared on a label (December 2012), https://ec.europa.eu/food/sites/food/files/safety/docs/labelling_nutrition-vitamins_minerals-guidance_tolerances_1212_en.pdf Documents are available in all EU official languages	
Austria	The Austrian Ministry of Health published Q&A on interpretation of the FIC Regulation, https://www.bmgf.gv.at/home/LMIV	
Belgium	In May 2016, The Federal Public Service (FPS) Health, Food Chain Safety and Environment published an FIC explanatory document (in French and Dutch): https://www.health.belgium.be/nl/levensmiddelen-die-met-het-oog-op-de-onmiddellijke-verkoop-worden-voorverpakt FPS also published (in French and Dutch) Q&A on interpretation of the FIC Regulation (March 2017) https://www.health.belgium.be/nl/faq-inzake-de-etikettering-van-levensmiddelen It has also summarised, in a flowchart, the rules on content and presentation of the declaration of nutritional values (in Dutch), https://www.health.belgium.be/sites/default/files/uploads/fields/fpshealth_theme_file/flowchart_nl.pdf	In November 2014, the Centre for Agricultural Products with the support of Vlaamish authorities published *A Guideline on Application of FIC Regulation in Labelling of Agricultural and Regional Products*, http://www.steunpunthoeveproducten.be/Portals/11/Etikettering%20van%20hoeve.pdf In December 2016, Fevia, the Federation of the Food and Drink Industry in Belgium, published a 3-minute video about nutrient labelling under FIC aimed at consumers, https://www.fevia.be/fr/presse/letiquetage-nutritionnel-devient-obligatoire-pour-tous-les-aliments-preemballes
Czech Republic	The Czech Association of Food Producers, in cooperation with the Legislation Department of the Czech Ministry of Agriculture, prepared Q&A (January 2015), http://www.foodnet.cz/polozka/?jmeno=Aktualizov%C3%A1no+-+Elektronick%C3%A1+p%C5%99%C3%ADru%C4%8Dka+PK+%C4%8CR+k+na%C5%99%C3%ADzen%C3%AD+%28EU%29+%C4%8D.+1169%2F2011+o+poskytov%C3%A1n%C3%AD+informac%C3%AD+o+potravin%C3%A1ch+spot%C5%99ebitel%C5%AFm&id=37344	

Table 3.2 (*Continued*)

Country	Government	Industry
	In 2013, the Czech Ministry of Agriculture (Czech MoA) published a guideline on the FIC Regulation for FBOs, which incorporates the text of FIC, COM Q&A and the TOLERANCES document, http://www.bezpecnostpotravin.cz/UserFiles/publikace/2014_Prirucka_pro_PPP_final_MZe.pdf In 2018, an updated edition of this guideline was published by the Czech MoA, http://www.bezpecnostpotravin.cz/UserFiles/Koubova%201/PPPPP%20ZELEN%C3%81.pdf	In June 2013, the Czech Association of Food Producers (CAFP) published a guideline on food labelling for consumers, http://www.foodnet.cz/polozka/?jmeno=Pr%C5%AFvodce+nov%C3%BDmi+pravidly+pro+ozna%C4%8Dov%C3%A1n%C3%AD+potravin+-+%22Ozna%C4%8Dov%C3%A1n%C3%AD+potravin+-+Mnoho+cenn%C3%BDch+informac%C3%AD+pro+spot%C5%99ebitele+%3F%22&id=36290 CAFP also published a flow chart on the labelling of vitamins and minerals, http://www.svazpekaru.cz/attachments/478_Postup%20p%C5%99i%20ozna%C4%8Dov%C3%A1n%C3%AD%20vitamin%C5%AF%20a%20ML.pdf
Denmark	The Danish Veterinary and Food Administration (2016) published a guideline on the FIC Regulation, including examples of mandatory and voluntary nutrition labelling. A list of standard/reference analytical methods used for nutrient labelling controls is also included, https://www.foedevarestyrelsen.dk/Selvbetjening/Vejledninger/naeringsdeklarationsvejledning/Documents/1.%20Vejledning%20om%20N%C3%A6ringsdeklaration.pdf	
Estonia	In its nutrition labelling guide (2016), the Estonian Veterinary Authority present examples of mandatory and voluntary labelling, http://www.vet.agri.ee/static/files/1454.Toitumisalane%20teave.pdf	
France	In December 2016, the General Directorate for Competition Policy, Consumer Affairs and Fraud Control published a guideline on mandatory and voluntary nutrition labelling, http://www.economie.gouv.fr/dgccrf/etiquetage-des-denrees-alimentaires-nouvelles-regles-europeennes	The French Association of Food Producers and Federation of Retailers published Q&A on interpretation of the FIC Regulation, including examples of mandatory and voluntary nutrition labelling (March 2013),

	In April 2019, Public Health France published a special web page on usage of "Nutri-Score" logo, which was accepted by the European Commission as a front-of-package scheme in 2017, https://www.santepubliquefrance.fr/determinants-de-sante/nutrition-et-activite-physique/articles/nutri-score	http://critt-iaa-paca.com/wp-content/uploads/2015/02/ANIA-FCD-questions-reponses-ANIA-FCD-sur-reglement-INCO.pdf In 2015, the Regional Association of Food Business from the region Occitane published a guideline on how operators can restore confidence in foods via labelling, http://www.agroalimentaire-lr.com/sites/aria.choosit.eu/files/fichiers%20personnels/Anne%20GERARD/164_LRIA_GuidePratique_etiquetage.pdf
Lithuania	In 2015, the State Food Veterinary Service published guidance in a Q&A format on labelling for prepacked foods that highlights new rules and includes a checklist to evaluate compliance with the FIC Regulation, https://vmvt.lt/sites/default/files/fasuotu_maisto_produktu_zenklinimo_gaires_02-13_0.pdf	
Ireland	In July 2012, the Food Safety Authority of Ireland published *Overview of Changes to Food Labelling Introduced under the New Food Information Regulation*, https://www.fsai.ie/Overviewofchangestofoodlabellingintroducedunderthenewfoodinformationregulation.html It also provides an online guideline: Food Information – FIC (Labelling), with eLearning modules on nutrition labelling, food information for prepacked foods and a summary of nutrition labelling requirements, https://www.fsai.ie/legislation/food_legislation/food_information_fic/food_information-fic.html and published a *Guide to Food Law for Artisan/Small Food Producers Starting a New Business* (2016) that helps to explain aspects of food law relevant to food businesses, https://www.fsai.ie/guidetofoodlawforartisanandsmallfoodproducers.html	In 2013, Food and Drink Industry Ireland published *Food and drink labelling: a practical guide for industry* which provides information about changes and presents a checklist for evaluation of compliance with FIC Regulations, https://www.webpackaging.com/Up/Comp/1450/11122419/11122421-IWCOSERB/f/Food%20and%20Drink%20LabellingGuide.pdf
Italy	The Ministry of Health has published an information sheet on what companies need to know about food labelling, http://www.salute.gov.it/imgs/C_17_opuscoliPoster_215_allegato.pdf	

Table 3.2 (*Continued*)

Country	Government	Industry
Luxembourg	In 2014, the Ministry of Health published a guide on food labelling for food producers in Luxembourg, with highlighted changes and exceptions and explanations of nutrition labelling, https://www.horesca.lu/uploads/files/etiquetage%20des%20denrees%20alimentaires.pdf	
Malta	In August 2013, the Environmental Health Directorate published *A focus on the labelling of food products*, https://health.gov.mt/en/environmental/Documents/Publications-Env-Health/7focus201308_labelling_en.pdf	
Netherlands	The Netherlands Food and Consumer Product Safety Authority has published guidance on FIC, https://www.nvwa.nl/onderwerpen/etikettering-van-levensmiddelen	In April 2015, FNLI (Dutch Food Industry Federation) published a factsheet on rules for nutrition labelling according to the FIC Regulation, https://www.fnli.nl/wp-content/uploads/2015/01/2015-04-v2-FNLI-Factsheet-regels-voor-etikettering-in-de-levensmiddelenindustrie-def-dec-2016-update.pdf
Poland	In 2015, the Advisory Agricultural Centre in Radom published a guideline on the requirements of food labelling, http://www.cdr.gov.pl/images/wydawnictwa/2015/2015-WYMAGANIA-W-ZAKRESIE-ZNAKOWANIA-PRODUKTOW-SPOZYWCZYCH.pdf In August 2016, the Agricultural and Food Inspection Authority published a guideline on the application of the FIC Regulation, which was updated in 2017, http://www.ijhars.gov.pl/pliki/A-pliki-z-glownego-katalogu/ethernet/2017/czerwiec/BOL/przewodnik%20do%20rozporzadzenia%201169%202011%20-%20czerwiec%202017.pdf	
Portugal	In March 2017, the Portuguese National Authority for Animal Health and the Portuguese Nutrition Society published a guide to food labelling choices, aimed primarily at consumers, http://www.apn.org.pt/documentos/ebooks/Ebook_Final_Rotulagem.pdf	In October 2013, FIPA (Portuguese Federation of Agricultural and Food Industries) and APED (Portuguese Federation of Retailers) published information for consumers on the application of the FIC Regulation, with a flow chart for rules on mandatory and voluntary nutrition labelling, http://www.fipa.pt/uploads/fotos_artigos/files/GuiaRegulamento.pdf

Romania	The Romanian Consumer Protection Authority published a Romanian version of the guidance prepared by EuroCommerce & FoodDrinkEurope (see below), http://www.anpc.gov.ro/anpcftp/interes_public/ghid_150226.pdf	
Slovakia	The State Veterinary and Food Administration published Q&A, https://www.svps.sk/dokumenty/potraviny/Otazky_odpovede_1.pdf	
Sweden	Guidance from the National Food Administration (April 2014) is available at https://www.livsmedelsverket.se/globalassets/om-oss/lagstiftning/livsmedelsinfo-till-konsum—markning/info-forordning-1169-2011-rev-2.pdf	
UK	In November 2012, the Department for Environment Food & Rural Affairs (DEFRA) produced *The Food Information Regulations 2013: Guide to compliance*, https://www.gov.uk/government/uploads/system/uploads/attachment_data/file/82663/consult-fic-guidance-20121116.pdf In 2014, Food Standards Agency Northern Ireland published a Food Standards Training Manual, *A Reference to Food Composition and Labelling Legislation*, http://niopa.qub.ac.uk/bitstream/NIOPA/2425/1/foodcompositionandlabelling_0.pdf In December 2014, the Food Standards Agency published *Food Information Regulations 2014: Summary guidance for food business operators and enforcement officers in Scotland, Wales and Northern Ireland*, https://www.food.gov.uk/sites/default/files/media/document/fir-guidance2014.pdf. and a summary leaflet: Nutrition Labelling Information for Food Businesses, https://www.food.gov.uk/sites/default/files/media/document/nutritionlabellinginformationleaflet.pdf In October 2015, the Department of Health published comprehensive information in a *Nutrition Legislation Information Sheet*, with extensive links to relevant authorities, trade associations, law documents and guidelines, https://www.reading.ac.uk/foodlaw/pdf/uk-15019-Nutrition-Legislation-Info.pdf In November 2016, the Department of Health updated its *Guide to creating a front of pack (FoP) nutrition label for pre-packed products sold through retail outlets*, https://www.gov.uk/government/uploads/system/uploads/attachment_data/file/566251/FoP_Nutrition_labelling_UK_guidance.pdf	In 2013, RT Coopers Solicitors published *Food Labelling: Presentation and Advertising of Foodstuffs: Nutrition Labelling: Provision of Food Information to Consumers*, http://www.rtcoopers.com/userfiles/file/Food_Labelling_Oct13(4).pdf The University Caterers Organisation (TUCO) published a *Nutritional Labelling Guidance Pack*, http://www.tuco.ac.uk/images/TUCO_Nutritional_Labelling_Guidence_Pack_2016.pdf

Table 3.2 (*Continued*)

Country	Government	Industry
	In March 2017, the Department of Health published its technical guidance on nutrition labelling, with a flowchart providing informal, non-statutory advice for businesses on the nutrition-related requirements, https://www.gov.uk/government/publications/technical-guidance-on-nutrition-labelling In November 2017, DEFRA and the Food Standards Agency updated its guidance on Food labelling: giving food information to consumers, https://www.gov.uk/guidance/food-labelling-giving-food-information-to-consumers The Food Safety Agency of Northern Ireland (FSANI) published a leaflet on nutrition labelling requirements, https://infograph.venngage.com/p/149934/nutritional-labelling-leaflet-fsani The Food Standards Agency, DEFRA and the Department of Health developed an e-learning course on food labelling including a nutrition labelling module, http://labellingtraining.food.gov.uk/ The National Health Service published a general introduction to food labels, https://www.nhs.uk/live-well/eat-well/how-to-read-food-labels/	
FoodDrinkEurope and EuroCommence		In September 2013, FoodDrinkEurope and EuroCommence published *Guidance on the Provision of Food Information to Consumers*, http://www.fooddrinkeurope.eu/uploads/publications_documents/FDE_Guidance_WEB.pdf

Spirits Europe		In December 2014, Spirits Europe published *Guidance for the Implementation of Food Information to Consumer Regulation (FIC)*, which also contained information about the specifics of nutritional labelling of spirits. http://www.spiritsnl.nl/downloads/gp.ci-024-2014-fic-guidance-v4-dec.2014-.pdf
GS1		In May 2016, GS1 published its *Manual: How to comply with the EU 1169/2011 regulation via the CDB (Central Data Bank)*, https://www.gs1belu.org/sites/gsonebelu/files/01072015_jl_rg_nc_how_to_comply_to_1169_via_the_cdb_-_mai_2016_mjr3_en.pdf GS1 also published various guidelines on the implementation of the FIC Regulation, http://www.gs1.eu/activity/eu-1169-regulation-on-food-labeling#publications Materials on implementation of the FIC Regulation are also available in languages of the countries where GS1 operates.

published on 31 January 2013 in the 24 EU official languages. This document had a particularly important role during the transition period for adoption of FIC by FBOs (13 December 2014 to 12 December 2016).[14] On 31 May 2018, the EC adopted a Commission Notice on these Q&A[15] that provided answers to additional questions concerning application of FIC, which were raised after entry into force; the Notice was drafted in collaboration with EU MS.

A similar Q&A approach has been used by some national trade associations, *e.g.* French Association of Food Producers and the Federation of Trade and Distribution Companies[20] and the Czech Association of Food Producers, in cooperation with the legislation department of the Czech Ministry of Agriculture and the Czech Agriculture and Food Inspection Authority.[21] Governmental departments have also provided Q&A, *e.g.* Belgian Federal Public Service,[22] the UK Food Standards Agency[23] and Slovak State Veterinary Office.[24]

3.3.1.2 Summary of Nutrition Legislation

Comprehensive information on nutrition legislation was published by the UK Department of Health in 2014, to "help food businesses comply with nutrition legislation. It may also be of use to others with an interest in the legislation, such as food law enforcement officers and trade associations".[25] Although the information sheet gave only a brief introduction to FIC, the main significance of this information was as a navigator, linking text to the legislation and extensive supporting information from authorities, trade associations, law documents and other guidelines. In Ireland, similar guidance was published for artisan/small food producers starting a new business.[26]

3.3.1.3 Approaches to Explanation and Examples of FIC Implementation

A common approach in guidelines for (FIC) food labelling has been a facilitated explanation of the individual provisions, including reference to FIC articles and national food laws, although there are different ideas about what constitute the most important issues:

- Indication of nutrition labelling prior to and after FIC[27–29]
- Checklist of mandatory and voluntary particulars of food and nutrition labelling[27,29]
- Highlighted mandatory and supplementary particulars[27,28]
- Examples of how mandatory and voluntary nutrient labelling might be combined[30,31]
- Visual indication of the most important particulars on food packaging, *e.g.* using a chart of folded or unfolded box[32]
- Flyers summarising the channels and messages[33–35]
- Decision-making flow charts to help navigate mandatory and voluntary particulars[36–38]

3.3.1.4 Additional Guidelines for Voluntary Labelling

FIC allows other forms of information besides text, such as graphics (symbols) for voluntary front-of-package labelling (FoP) (ref. 1, Article 35(1)).

In the UK, colour coding (red, amber and green) is used for fat, saturates, sugars and salt, and step-by-step guidelines, including how to calculate reference intakes, criteria for colour coding, information for consumers, and worked examples, have been developed by the national authority.[39] The UK system was introduced in 2013.[40]

The French government introduced the Nutri-Score scheme in 2017.[41,42] The scheme ranks nutritional value of products, based on the score, which considers quantities of specific nutrients to limit or to encourage consumption.[40] A single code label uses a palette of five colours (from green to red) and letters A to E.[5,43] Guidelines[43] and frequently asked questions[44] about the criteria and use of the scheme have been published by the French authorities.

The Nordic Keyhole scheme was introduced in 1989. The label is also used in Denmark, Norway and Iceland as well as in Lithuania and Macedonia. Its primary aim is to help consumers to identify the healthiest choice in each of 33 product categories and encourage manufacturers to develop healthier products.[40,45] Conditions for displaying this label consider content of salt, sugar, fat, saturated fat, fibre and energy content, limits for which are defined for each product category.[45]

FIC required the EC to adopt a report on additional forms of expression and presentation for nutrition declarations and harmonisation in this field by 2017 (ref. 1, Article 35(5)). This was postponed until the end of 2018, as experience with governmental and other front-of-pack (FoP) systems is to be included in the Joint Research Centre (JRC) literature review and the subsequent European Commission report; both are expected to be published in 2019.[40] In general, FoP issues have long been a subject of intense debate. Introduction of the French FoP scheme was described by FoodDrinkEurope – European Association of Food Producers – as an example of: "proliferation of national schemes that hamper consumer understanding and would be a barrier to the single market".[46] In April 2018, the EC organised a joint meeting of EU MS representatives and interested parties focused on technical aspects related to FoP, including discussion on existing schemes and sharing information, with the aim of achieving consensus on harmonisation of FoP. A second meeting was held in June 2018.[40] However, whilst there is agreement on the need for harmonisation of FoP schemes in Europe, finding a consensus remains elusive.[5]

3.3.1.5 Online Guidelines and e-Learning Courses

Some authorities have introduced e-learning to explain nutrition labelling. The Food Safety Authority of Ireland (FSAI, www.fsai.ie) has published a 20-minute e-learning module[47] that guides users though nutrition information rules, explains how and when these apply to food businesses, and provides an opportunity to check compliance of nutrition information under FIC. The nutrition labelling rules are presented using compliance checks that highlight the most common problems and mistakes in presentation of nutrient declarations. This e-learning module provides a defined package of information about nutrition labelling.[47]

Similar online guidance on food and nutrient labelling[48] has been published by the UK Food Standards Agency (FSA, www.food.gov.uk) in partnership with the Department for Environment, Food & Rural Affairs (DEFRA).[48] It is comprised of 13 modules, including information about nutrition labelling regulated by FIC and other related regulations on nutrition and health claims,[12] labelling of foodstuffs for particular nutritional uses, food supplements and fortified foods, and national measures. Self-assessment in the form of a quiz is available for each module and users can apply for a Continuing Professional Development certificate on completion. The e-learning has a readily accessible glossary of terms and is presented such that, overall, performance meets the primary goal, which is to improve general understanding of current food labelling legislation.

3.3.2 Guidelines on Obtaining Nutrient Values

FIC permits three methods for obtaining nutrition values used in a nutrient declaration. Accordingly, data shall be average values based on (ref. 1, Article 31(4)):

1. Manufacturer's (chemical) analysis of the food.
2. Calculation from the known or actual average values of the ingredients used.
3. Calculation from generally established and accepted data.

Values must comply with FIC legislation with respect to procedures for evaluating tolerances in nutrition declaration guidance documents from competent authorities. In other words, manufacturers are legally responsible for the values declared and must ensure values are within permitted tolerances.[16] Tolerances are based on variations in food products and define acceptable differences between nutrient values declared on a label and values established by monitoring authorities. Variations arise from naturally occurring differences in food products, depending on intrinsic (*e.g.* variety, age and breed of animal) and extrinsic factors (*e.g.* climate and weather, season, feed, processing) as well as average chemically analysed values, published in national food composition tables, compared with individual varieties. FIC also defines rounding rules for nutrition declarations.[16]

Table 3.3 presents data on results from inspection of nutrition declaration in the Czech Republic during 2017.[49] Evaluation of tolerances was performed by the Czech Agricultural and Food Inspection Authority in compliance with the guidance documents for competent authorities.[16] Non-compliance was discussed with producers to determine reasons for incorrect

Table 3.3 Results of inspections focused on nutrient labelling under FIC conducted by the Czech Agricultural and Food Inspection Authority in 2017.

	Samples		
Nutrient	Analysed	Not compliant	% Not compliant
Proteins	833	21	2.5
Fat	937	15	1.6
Carbohydrates/sugars	880	14	1.6
Vitamins	420	20	4.8
Vitamin C	143	12	8.4
Energy value	23	1	4.3
Salt	156	24	15.4

values and explain correct procedures for nutrient declaration of prepacked foods. Results indicated a higher percentage of incorrect values for salt than other mandatory nutrients. Inspections will continue, with particular attention given to salt content.

3.3.2.1 Manufacturers' Analyses

Manufacturers can obtain analytical data in-house or through contracted analytical laboratories that use internationally standardised methods. Food analysis is expensive and time-consuming, but it is the preferred method for obtaining food composition data. However, it is not always appropriate for SME FBOs with only a few product lines. Furthermore, neither FIC nor the majority of national guidelines (Table 3.2) recommend methods of analysis or advice about understanding results from such chemical analyses. The only exception is the Danish Veterinary and Food Administration (www.foedevarestyrelsen.dk), which includes a list of standardised or referenced methods for obtaining nutrition values.[30]

3.3.2.2 Calculation of Values

FIC does not necessarily require laboratory analysis.[1] Instead, manufacturers can use calculations, based on known or actual average values of the ingredients, *e.g.* published data or data compiled by analysis. Generally, these types of data track composition over months and years, enabling end users to be reasonably confident in proximate values.[50]

Manufacturers can also calculate values, based on generally established and accepted data, *i.e.* national food composition tables. Some authorities recommend a national food composition database/tables for this purpose. For example, in the UK, the official *Technical guidance on nutrition labelling* states[51] that: "In the UK, 'generally established and accepted data' can be found in book form in McCance & Widdowson's The Composition of Foods or online in McCance & Widdowson's The Composition of Foods integrated dataset (CoF IDS) at: https://www.gov.uk/government/publications/composition-of-foods-integrated-dataset-cofid."

Calculation procedures (hereafter recipe calculation) are not defined by FIC, which FBOs need to be aware of potential issues and sources of errors. Calculations are based on amounts of ingredients in a recipe and nutrient composition of these ingredients, and values need to be adjusted for weight (*e.g.* loss or gain of water) and nutrient content (*e.g.* loss through cooking) using yield[52,53] and retention factors,[53,54] respectively. Lack of experience with recipe calculation and food composition data can easily result in incorrect values, even if FBOs use commercial software.[55] There are different methods for recipe calculations in Europe, as revealed by an inventory of calculation methods[56] performed within the EU FP6 EuroFIR (European Food Information Resource Network of Excellence, 2005–2010), and most software does not declare the approach used or source of established and accepted data.

In fact, the quality of nutritional labelling software (NLS) available on the market is highly variable, meaning results could mislead both FBOs and consumers. There is an urgent need to ensure NLS is authoritative and transparent, and generates accurate information for consumers and FBOs, which are responsible for published values, regardless of who calculated them.

Providing FBOs with guidelines on recipe calculation is essential for understanding the process and identifying potential sources of errors, as well as determining when chemical analysis is more appropriate. To that end, EuroFIR AISBL (www.eurofir.org) has proposed a harmonised recipe calculation procedure,[57] where yield factors are applied at the recipe level (*i.e.* to all ingredients), and standardised retention factors for vitamins and minerals used for each ingredient,[54] which have been adopted by a few companies offering recipe calculation services (*e.g.* Etiketomat, www.etiketomat.cz; Vyraex, www.vyraex.cz; and Nutritics, www.nutritics.com (IR)).

3.3.2.2.1 EuroFIR Guideline on Recipe Calculation Procedure for FBOs. Recipe calculation guidelines developed by EuroFIR AISBL, *How to Calculate Nutrient Content of Foods: A Guideline for Food Business Operators*, are primarily a tool for FBOs to fulfil their legal obligation for nutrition declarations in the EU.[58] The guidelines are based on the harmonised recipe calculation procedure proposed by EuroFIR. The independence and reputation of EuroFIR, as a guarantor for these guidelines, is important for implementation at national levels. This process starts with translation into national language(s), prior to negotiation with national authorities and promotion amongst target users.

This approach was applied first in the Czech Republic and then Croatia and, beyond the EU, in Norway (Figure 3.1); Slovakia also commenced implementation in 2017.

How to Calculate Nutrient Content of Foods: A Guideline for Food Business Operators is organised into two parts. *Part I General description* explains recipe calculation procedures in 10 steps, and sources of food composition data and standardised retention factors are provided. Limitations in the calculations are explained and the importance of validation (of calculated values) with data from chemical analysis is stressed. Responsibilities of

Figure 3.1 Different language versions of the EuroFIR Recipe Calculation Guideline.

FBOs in deciding on the suitability of calculation for a product or type of product is also emphasised.

The guidelines focus on calculation of nutrient content for mandatory declarations, fibre and sodium without application of retention factors. Calculation of vitamins and minerals using retention factors is also considered, although these are not always declared because many foods do not meet the "significant amount" threshold (ref. 1, Annex XIII), the only exceptions being EU MS that have made vitamin and mineral declarations mandatory. Chemical analysis of foods is discussed, as the preferred method for nutrition declaration on products bearing nutrition or health claims[59] or declaring content of vitamins and minerals.

Part II includes practical examples and is structured as individual Excel worksheets (Figure 3.2) demonstrating documentation of a recipe, calculation without and with retention factors, comparison of calculated and analysed values considering compliance with tolerances for food labelling, calculation of water content and a check of total proximates.

The guidelines could be considered the first step in future harmonisation of recipe calculations across Europe. To achieve this, however, three prerequisites need to be fulfilled:[60]

1. Validation of calculated data by chemical analysis, which will allow the suitability of calculation procedures for different foods and food processing to be determined.
2. Training of all user groups to improve understanding and use of recipe calculation.
3. Implementation of the guideline at the national level in other EU MS.

3.3.2.2.2 Lithuanian Recipe Calculation Guidelines. The Lithuanian Veterinary Authority (https://vmvt.lt) has published recipe calculation guidelines, developed by the Kaunas Technical University in 2014, together with a list of databases for sourcing food composition data. The guidelines have been approved by the national Ministry of Health, which also approved a list of weight changes (yield factors) during processing.[61]

The recipe calculation procedure is the same as that recommended by EuroFIR, with examples and limitations explained in detail. The need to validate calculated values by chemical analysis is also emphasised. Chemical analysis is also recommenced as the preferred method for obtaining nutrient values.

3.3.2.2.3 Non-specialist User *vs.* Expert Assisted Recipe Calculation. Recipe calculation is a legally acceptable and cost-effective alternative to chemical analysis, especially for SME FBOs. However, it is not a simple procedure and users must consider specific challenges[55] in producing average values that best represent the amounts of nutrient a food contains. It is necessary to find appropriate nutrient values for each ingredient, adjust values in compliance with food labelling regulations, convert

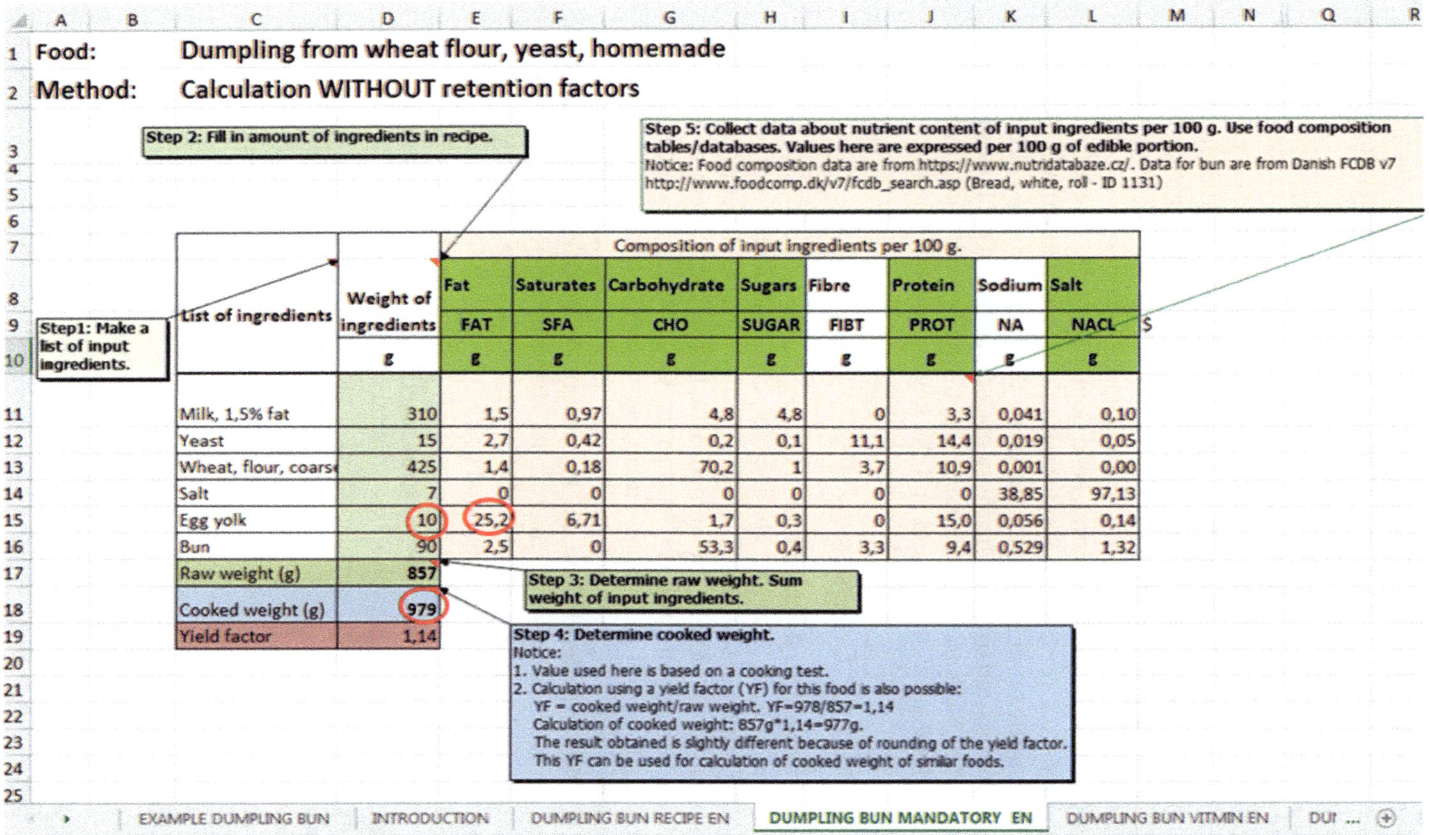

List of ingredients	Weight of ingredients	Fat	Saturates	Carbohydrate	Sugars	Fibre	Protein	Sodium	Salt
		FAT	SFA	CHO	SUGAR	FIBT	PROT	NA	NACL
	g	g	g	g	g	g	g	g	g
Milk, 1,5% fat	310	1,5	0,97	4,8	4,8	0	3,3	0,041	0,10
Yeast	15	2,7	0,42	0,2	0,1	11,1	14,4	0,019	0,05
Wheat, flour, coars	425	1,4	0,18	70,2	1	3,7	10,9	0,001	0,00
Salt	7	0	0	0	0	0	0	38,85	97,13
Egg yolk	10	25,2	6,71	1,7	0,3	0	15,0	0,056	0,14
Bun	90	2,5	0	53,3	0,4	3,3	9,4	0,529	1,32
Raw weight (g)	857								
Cooked weight (g)	979								
Yield factor	1,14								

Figure 3.2 The EuroFIR Recipe Calculation Guideline: Part II, Practical examples – calculation of nutrients for mandatory nutrient declaration: an example of a table for collection of input data and recommendations for users. Mandatory nutrients are marked green.

units and household measures to weights, consider weight changes and nutrient losses during processing (cooking), and evaluate the impact of complex recipes (*e.g.* cooked ingredients combined with raw ingredients, products that are cooked twice [*e.g.* fried and baked]).

Undertaking recipe calculation is challenging, even for experienced individuals. It should not be assumed that non-specialists (*i.e.* most FBOs) will be able to produce accurate information, even if they use recipe calculation software. Accuracy could be improved through training,[55] either face-to-face or online. Another option could be to improve the performance of recipe calculations through so-called "assisted" recipe calculation, supported by national experts in the field, *e.g.* compilers of national food composition databases, as provided by the Slovak Food Composition Databank.[62] In the UK, a similar service is provided by Food Databanks National Capability (FDNC) at the Quadram Institute Bioscience (formerly the Institute of Food Research), which maintains and updates the Nutrient Composition of UK Foods database and publishes *McCance & Widdowson's The Composition of Foods* in collaboration with Public Health England and the Royal Society of Chemistry.[63]

3.4 Training Courses

Training courses on the application of FIC, organised by authorities and trade associations, are an effective channel for explaining rules and providing practical examples of labelling (both food and nutritional). They are organised on national or EU levels and aimed at a broad community of learners, including FBOs, authorities and inspectors. Generally, they are delivered as face-to-face instructor-led training, taking advantage of real-time discussions, immediate feedback, learner networking and direct communication of learners with experienced tutors. Presentations from courses, published online, can serve as a useful reference for practical applications of food law and, in particular, provisions of FIC.[64–66] In this section, training activities around application of FIC at national and EU levels are considered briefly.

The majority of courses have focused on general application of FIC, including provisions on layout of nutrient labelling, and provide summaries of the content and presentation of nutritional declarations. There is a lack of training courses on the calculation of nutritional information, including online resources and use of specialised software for experienced or inexperienced food industry personnel. As described above, recipe calculation is not an easy issue and training courses should take into consideration, as a minimum, the following:

- Food composition, including data sources, data compilation and implications when using data for food labelling.
- Calculation methods and factors that affect data quality.
- Specific calculation tools, both manual and software based.
- Gap filling and specific ingredient considerations, including variations, tolerances and errors.

- Data validation and data quality.
- Transformation of calculated nutritional information into compliant labels.

Inclusion of this framework could help FBOs to gain knowledge and skills in recipe calculation, contribute to their understanding of the procedures, and help ensure accurate nutrient values are published on labels not only to inform consumer choice but also protect FBOs from repercussions arising from non-compliance, because "compliance with new legislation can impact future success".[67]

Examples of selected instructor-led training (ILT) courses are given in Table 3.4.

3.5 Consumer Education

The main target group for nutrition labelling are consumers, the aims being to provide information about the current legislation framework (*i.e.* FIC) and explain the nutritional information with the goal of improving consumer health through more informed choices.[5,69–72] In order for consumers to act upon the nutrition information provided, they need to be – at the very least – aware of the regulatory measures and their main provisions. Thus, the introduction of FIC was used concomitantly as an opportunity to inform consumers about the new rules using a variety of information channels and formats, *e.g.* newsletters published by authorities,[73] presentations in the media (television, newspapers, journals), guidelines[74] and leaflets.[75] One interesting approach was published by the Belgian Federal Public Service Health, Food Chain Safety and Environment in cooperation with Fevia, the Federation of the Food and Drink Industry in Belgium. An approximately 3-minute video explains briefly content and position of mandatory and voluntary nutrition declarations under FIC and demonstrates sources of nutrients. This information is presented in a well-structured, inclusive and informative form whilst still providing the consumer with most important aspects of the legislation.[76]

3.6 Conclusion

The reasons for EU MS governments and others developing regulations and standards on nutrition labelling are summarised by Hawkes[77] and included the need to:

- provide a standard format for labelling nutrients, preventing continued use of a potentially confusing multitude of different formats by different food companies;
- ensure that FBOs include less desirable nutrients (*e.g.* saturated fats) as well as nutrients that are perceived positively (*e.g.* vitamins);
- demonstrate that nutrition claims are honest and truthful;
- ensure that nutrition labelling is not false, misleading or deceptive;

Table 3.4 Examples of instructor-led training (ILT) courses on provisions of FIC concerning nutrient labelling and obtaining food composition values for nutrient declaration.

Course subject	Organisation level	Country	Organiser	Description of course (all links accessed November 2018)	Target learners
Content and expression of nutritional declarations	EU	EU	European Commission	Better Training for Safer Food (BTSF) is a training initiative covering food and feed law, animal health and welfare and plant health rules. Targeted tailor-made training in various formats: https://ec.europa.eu/food/safety/btsf_en	Train the trainers principle. The courses are delivered in EU and non-EU countries, targeting the staff of competent authorities dealing with official controls from EU and selected non-EU countries. The participants should disseminate the knowledge acquired from the training amongst their colleagues in their home countries.
			Consumers, Health, Agriculture and Food Executive Agency (formerly CHAFEA)	Food composition and information training includes EU rules on general food labelling and claims as well as on specific categories of foods such as food supplements, foods with added vitamins and minerals and foods for specific groups of the population: https://foodinfo-europe.com/index.php/projects/fci#contact	
	National	Czech Republic	Department for Food Legislation of the Czech Ministry of Agriculture, Czech Agricultural and Food Inspection Authority, and the Czech Association of Food Producers	A series of ILT courses organised prior to FIC and during the transitional period. Participants had the chance to obtain information directly from policy-makers, inspectors and representatives of food manufacturers, and be informed about the position of authorities regarding the labelling rules: http://www.foodnet.cz/slozka/?jmeno=Semin%C3%A1%C5%99e&id=109&archiv_rok=2014	FBOs
			Department for Food Legislation of the Czech Ministry of Agriculture	A specialised course on provisions of FIC regarding nutrient labelling aimed at harmonisation of inspection interpretations.[68]	Food inspectors

		Commercial agencies	One-day courses are a common ILT format. They are usually focused on the main provisions of FIC including nutrient labelling information of interest to different user groups, including enforcement practice of the Czech food inspection authorities: https://www.vox.cz/potravinarska-legislativa/kurz-oznacovani-potravin-pro-pokrocile-vcetne-praktickych-prikladu-0001.htm http://www.bohemiacert.cz/training/ https://www.kursy.cz/poskytovani-informaci-o-potravinach-productsmpo/	FBOs, quality, product and marketing managers, import officers, translators, managers of e-shops
National	Belgium	The Federation of the Food and Drink Industry in Belgium (Fevia)	Fevia collaborates closely with 27 branch organisations that offer nutrition labelling advice specific to their sector: https://www.fevia.be/nl/leden-en-sectoren	FBOs
		Alimento groups	Sectoral services managed by social partners offering training for employers and employees in the food industry, including nutrition labelling: https://www.alimento.be/nl/training/?page=3	FBOs
National	Netherlands	The Dutch Food Industry Federation (FNLI)	FNLI is helping food companies to publish the correct nutritional information: https://www.fnli.nl/werkgebieden/veiligheid-regelgeving/	FBOs
		Etikettiren.nl	Commercial training courses intended for employees and managers who are responsible for preparing and checking nutrient labels in the Netherlands: https://etiketteren.nl/index.php/training-workshop	FBOs

Table 3.4 (*Continued*)

Course subject	Organisation level	Country	Organiser	Description of course (all links accessed November 2018)	Target learners
			Food Law Academy (FLA), a joint initiative of Wageningen Business School, European Institute for Food Law and Dijkstra Voermans	The FLA is a food law knowledge centre that collects and shares food legal knowhow through short courses, summer schools, in-company training or online teaching. Aimed at advancing the level of expertise on food law and regulatory compliance: https://www.food-law-academy.com	People with a technical background and lawyers working in the industry, consultancy, governmental or academic functions in the area of food regulatory affairs
	National	UK	Leatherhead Food Research	Its Legal Labels and EU Food Law course provides learners with comprehensive training on EU food labelling and other legislation applicable to food product compliance: https://www.leatherheadfood.com/event/legal-labels-and-eu-food-law-oct-2018/	FBOs, enforcement officers and other interested subjects
	National	Italy	Association of Pasta, Chocolate, Sweets, Confectionery Industry	A training day on food labelling targeted at particular food commodities: http://www.aidepi.it/en/news-engl/375-training-day-reg-eu-no-1169-2011-how-to-build-new-labels-focus-on-claims.html?jjj=1535611743926	FBOs
Obtaining nutritional values (*e.g.* recipe calculation)	EU	EU	NASCENT (grant no. 739922) project coordinated by EuroFIR	A series of training activities focused on knowledge of labelling regulations and knowledge of potential errors and different methods for nutritional information calculation/analysis was organised with the aim to "develop and extend nutritional labelling software and services for food manufacturers including SMEs, based on a pilot offering in the UK and Ireland (Nutritional Information Solutions). This	FBOs, especially SMEs

			exploits expertise within EuroFIR in terms of food composition data and standardisation of recipe calculation. The extended service will provide an accredited nutritional labelling service for food producers, using an existing software platform that goes beyond the automated services currently available." A certification scheme for nutritional labelling software is one of the outputs of the NASCENT project. http://www.eurofir.org/our-resources/current-projects/nascent/	
National	UK	Susan Church Nutrition	The Recipe Analysis: Maximising Accuracy training course is an example of a training course that offers practical and educational sessions designed to update and enhance knowledge and skills in recipe calculation: https://www.susanchurchnutrition.co.uk/recipe-analysis-training/	Dietitians, nutritionists, FBOs
National	Ireland	Nutritics	An example of training courses organised by a provider of nutritional labelling software. Advanced recipe calculation certification course and regular weekly webinars are organised to support software users: https://www.nutritics.com/p/news_Advanced-Recipe-Calculation-Certification-by-Nutritics	Nutritional labelling software users
National	Czech Republic	Department for Food Legislation of the Czech Ministry of Agriculture	A specialised course for inspectors at the Czech Agricultural and Food Inspection Authority on obtaining food composition values (both by chemical analysis and the recipe calculation method recommended by EuroFIR; see Section 3.2.2.1).[68]	Food inspectors

- encourage FBOs to apply sound nutrition principles in the formulation of foods;
- use formats that are effective and encourage consumers to make healthier choices;
- meet nutrition labelling requirements of other countries, thus facilitating export.

Once regulations and standards are in force, FBOs must comply with them or risk being prosecuted. Implementation of these labelling measures is costly[78] and so support for FBOs from national authorities is necessary to ensure accurate information is published, and this has been made available to some degree. However, FBOs need detailed information about nutrition labelling rules with examples, especially where exceptions or errors might readily arise. Thus, continued cooperation of authorities, regulatory and enforcement, and the food industry is necessary to ensure nutrition labels are correct and consumers sufficiently informed to use this information when making dietary choices.

Acknowledgements

Nutritional labelling software and claims: service, training and innovation offering to SMEs and industry in Europe (NASCENT).

This project has received funding from the European Union's Horizon 2020 research and innovation programme under grant agreement no. 739622.

References

1. European Parliament and the Council, Regulation (EU) No 1169/2011, On the provision of food information to the consumers, *Off. J. Eur. Communities: Legis.*, 2011, **L304**, 18–63.
2. The Council of the European Communities, Council Directive No 90/496/EEC, *Off. J. Eur. Communities: Legis.*, 2011, **L276**, 40–44.
3. M. J. B. van der Meulen, The structure of European Food Law, *Laws*, 2013, **2**, 69–98.
4. M. Ceasar, *Is Food Labelling the Best communication Tool for Protecting Consumers?,* 2016. https://blog.ifis.org/food-labelling-communication, (accessed November 2018).
5. European Food Information Council, *Global Update on Nutrition Labelling – Executive Summary*, Brussels, Belgium, 2018. https://www.eufic.org/images/uploads/healthy-living/Executive-Summary-GUNL-2018-V2.pdf, (accessed November 2018).
6. W. Kasapila and S. M. Shaarani, *Crit. Rev. Food Sci. Nutr.*, 2016, **56**(1), 56–64.
7. Codex Alimentarius International Food Standards, http://www.fao.org/fao-who-codexalimentarius/thematic-areas/nutrition-labelling/en/, (accessed November 2018).

8. ASEAN Food and Beverage Alliance, *Nutrition Labelling on Prepacked Food Impact on Trade in ASEAN*, 2018. http://afba.co/wp-content/uploads/2018/02/Nutrition-Labelling-on-Prepackaged-Food-Impact-on-Trade-in-ASEAN.pdf, (accessed November 2018).
9. 10th Seminar on Nutrition Labelling, Claims and Communication Strategies, 2018. http://ilsi.org/event/10th-nutrition-labeling/, (accessed November 2018).
10. EUR-Lex Access to European Union law, https://eur-lex.europa.eu/homepage.html, (accessed November 2018).
11. Micro-, small- and medium-sized enterprises: definition and scope, https://eur-lex.europa.eu/legal-content/EN/TXT/?uri=LEGISSUM%3An26026, (accessed November 2018).
12. Food – Food Safety – European Commission, https://ec.europa.eu/food/safety_en, (accessed November 2018).
13. Food information to consumers – legislation, https://ec.europa.eu/food/safety/labelling_nutrition/labelling_legislation_en, (accessed November 2018).
14. Commission's Health and Consumer Directorate General – DG SANCO. *Questions and Answers on the application of the Regulation (EU) No 1169/2011 on the provision of food information to consumers*, https://ec.europa.eu/food/sites/food/files/safety/docs/labelling_legislation_qanda_application_reg1169-2011_en.pdf, (accessed November 2018).
15. European Commission, Commission Notice on questions and answers on the application of Regulation (EU) No 1169/2011 of the European Parliament and of the Council on the provision of food information to consumers, *Off. J. Eur. Communities: Inf. Not.*, 2018, **C196**, 1–14.
16. European Commission, Guidance document for competent authorities, tolerances for the control of compliance of nutrient values declared on a label with EU legislation, 2012. https://ec.europa.eu/food/sites/food/files/safety/docs/labelling_nutrition-vitamins_minerals-guidance_tolerances_1212_en.pdf, (accessed November 2018).
17. Food Lawyers' Network Worldwide, EU-Food Information Regulation 1169/2011 Implementing Measures by Member States. Books on Demand, Norderstedt, 2015. https://www.kneppelhout.nl/media/rtf/11114916_buchblock.pdf, (accessed November 2018).
18. A. H. Meyer, National Implementing Regulation on the European Regulation No 1169/2011, Meyer rechtsanwalte, Munich, 2017. https://www.meyerlegal.de/pdf/newsletter-02-2017.pdf, (accessed November 2018).
19. N-Lex – Access to National law – choose your language, http://eur-lex.europa.eu/n-lex/, (accessed November 2018).
20. Association Nationale des Industries Alimentaires, Federation des entreprises du commerce et de la distribution, *Guide Questions-Reponses ANIA-FCD pour l'application du Reglement (UE) No 1169/2011 concernant l'information des consommateurs sur les denrées alimentaires*, 2013. http://critt-iaa-paca.com/wp-content/uploads/2015/02/ANIA-FCD-questions-reponses-ANIA-FCD-sur-reglement-INCO.pdf, (accessed November 2018).

21. M. Chylkova, N. Stepanek and D. Zima, Otazky a odpovedi k narizeni (EU) č. 1169/2011 o poskytovani informaci o potravinach spotrebitelum. Prakticky pruvodce novymi pravidly pro oznacovani potravin. *Potravinarska komora Ceske Republiky*, 2015. http://www.foodnet.cz/polozka/?jmeno=Aktualizov%C3%A1no+-+Elektronick%C3%A1+p%C5%99%C3%ADru%C4%8Dka+PK+%C4%8CR+k+na%C5%99%C3%ADzen%C3%AD+%28EU%29+%C4%8D.+1169%2F2011+o+poskytov%C3%A1n%C3%AD+informac%C3%AD+o+potravin%C3%A1ch+spot%C5%99ebitel%C5%AFm&id=37344, (accessed November 2018).
22. The Federal Public Service (FPS) Public Health, Food Chain Safety and Environment, *FAQ on the labelling of foodstuffs*. https://www.health.belgium.be/nl/faq-inzake-de-etikettering-van-levensmiddelen, (accessed November 2018).
23. Food Standards Agency, *Questions and Answers on the nutrition requirements contained in Regulation (EU) No 1169/2011 on the provision of food information to consumers (FIC)*. https://www.food.gov.uk/sites/default/files/media/document/qanda-nutritionalrequirementsof11692011.pdf, (accessed November 2018).
24. Otazky a odpovede o ulatnovani Nariadenia Europskeho parlamentu a Rady (EU) č. 1169/2011 v suvislosti s poskytovanim informacii o potravinach spotrebitelom. https://www.svps.sk/dokumenty/potraviny/Otazky_odpovede_1.pdf, (accessed November 2018).
25. Department of Health, *Nutrition Legislation Information Sheet*, 2014. https://www.reading.ac.uk/foodlaw/pdf/uk-15019-Nutrition-Legislation-Info.pdf, (accessed November 2018).
26. Food Safety Authority of Ireland, *Guide to Food Law for Artisan/Small Food Producers Starting a New Business*, 2016. https://www.fsai.ie/guidetofoodlawforartisanandsmallfoodproducers.html, (accessed November 2018).
27. Food and Drink Industry Ireland, *Food and drink labelling: a practical guide for industry*, 2013. https://www.webpackaging.com/Up/Comp/1450/11122419/11122421-IWCOSERB/f/Food%20and%20Drink%20LabellingGuide.pdf, (accessed November 2018).
28. Service de la sécurité alimentaire, *Étiquetage des denrees alimentares guide pour les enterprises du Luxembourg*, 2014. https://www.horesca.lu/uploads/files/etiquetage%20des%20denrees%20alimentaires.pdf, (accessed November 2018).
29. State Food Veterinary Service, *Fasuoto maisto productu ženklinimo garės*, 2015. https://vmvt.lt/sites/default/files/fasuotu_maisto_produktu_zenklinimo_gaires_02-13_0.pdf, (accessed September 2019).
30. Ministry of Environment and Food, Danish Veterinary and Food Administration, *Vejledning om næringsdeklaration (i henhold til forordning (EU) nr. 1169/2011 om fødevareinformation til forbrugerne)*, 2016. https://www.foedevarestyrelsen.dk/Selvbetjening/Vejledninger/naeringsdeklarationsvejledning/Documents/1.%20Vejledning%20om%20N%C3%A6ringsdeklaration.pdf, (accessed November 2018).

31. Estonian Veterinary Authority, *Toitumisalane teave*. https://vet.agri.ee/static/files/1454.Toitumisalane%20teave.pdf, (accessed November 2018).
32. Ministero della Salute, *Etichettatura degli alimenti. Cosa dobbiamo sapere*, 2015. http://www.salute.gov.it/imgs/C_17_opuscoliPoster_215_allegato.pdf, (accessed November 2018).
33. Nutrition Information under FIC, https://www.fsai.ie/training/nutrition_labelling/story_content/external_files/Summary.pdf, (accessed November 2018).
34. Nutritional labelling leaflet-FSANI. https://infograph.venngage.com/p/149934/nutritional-labelling-leaflet-fsani, (accessed November 2018).
35. Food Standards Agency, *Nutrition Labelling Information for Food Businesses*, https://www.food.gov.uk/sites/default/files/media/document/nutritionlabellinginformationleaflet.pdf, (accessed September 2019).
36. Annex: Summary of the EU FIC provisions governing expression of nutrition information per 100g/ml and per portion/consumption unit. https://www.gov.uk/government/uploads/system/uploads/attachment_data/file/207843/2012-10-24_Flow_chart_Expression_per_100g_and_per_portion.pdf, (accessed November 2018).
37. Potravinarska Komora Ceske Republiky, Jak pri oznacovani pristupovat k vitaminum a mineralnim latkam v potravinach. http://www.svazpekaru.cz/attachments/478_Postup%20p%C5%99i%20ozna%C4%8Dov%C3%A1n%C3%AD%20vitamin%C5%AF%20a%20ML.pdf, (accessed November 2018).
38. Samenvatting van de bepalingen betreffende de inhoud en uitdrukking van de voedingswaardevermelding. https://www.health.belgium.be/nl/samenvatting-van-de-bepalingen-betreffende-de-inhoud-en-uitdrukking-van-de-voedingswaardevermelding, (accessed November 2018).
39. Guide to creating a front of pack (FoP) nutrition label for pre-packed products sold through retail outlets. https://www.gov.uk/government/uploads/system/uploads/attachment_data/file/566251/FoP_Nutrition_labelling_UK_guidance.pdf, (accessed November 2018).
40. European Commission, *Minutes of the joint meeting on front-of-pack nutrition*, 2018. https://ec.europa.eu/food/expert-groups/ag-ap/adv-grp_fchaph/wg_2018_en, (accessed November 2018).
41. C. Julia, F. Etilé and S. Hercberg, *The Lancet Public Health*, 2018, **3**, 164.
42. European Commission, *Notification Detail*, 2017. http://ec.europa.eu/growth/tools-databases/tris/en/search/?trisaction=search.detail&year=2017&num=159, (accessed November 2018).
43. Santépublique France, *Usage regulation for the "Nutri-Score" logo*, 2018. https://www.santepubliquefrance.fr/content/download/3544/27772/file/Nutriscore_reglement_usage_EN_200218.pdf, (accessed November 2018).
44. Santépublique France, Nutri-Score frequently asked questions, 2018. https://www.santepubliquefrance.fr/content/download/3797/29487/file/QR_scientifique_technique_EN_190118.pdf, (accessed April 2018).

45. The Keyhole. https://www.livsmedelsverket.se/en/food-and-content/labelling/nyckelhalet, (accessed November 2018).
46. FoodDrinkEurope, *Statement on French "NutriScore" proposal*, 2017. https://www.fooddrinkeurope.eu/news/statement/statement-on-french-nutriscore-proposal/, (accessed 2018).
47. Nutrition Information under FIC, Regulation 1169/2011. https://www.fsai.ie/food_businesses/food_safety_training/online.html, (accessed November 2018).
48. Food labelling e-learning course. http://labellingtraining.food.gov.uk/, (accessed November 2018).
49. P. Cuhra, Prednaska – Vysledky kontrolni cinnosti Statni zemedelske a potravinarske inspekce za rok 2017, Seminar – Potravinarsky uterek, 26.6.2018, Praha, Czech Republic.
50. M. Howie, *BNF Nutr. Bull.*, 2015, **40**, 104–106.
51. Department of Health, *Technical guidance on nutrition labelling*, 2017. https://www.gov.uk/government/uploads/system/uploads/attachment_data/file/595961/Nutrition_Technical_Guidance.pdf (accessed November 2018).
52. A. Bognár and J. Piekarski, *J. Food Compos. Anal.*, 2000, **13**(4), 391–410.
53. L. Bergström, *Nutrient Losses and Gains in the Preparation of Foods*, Rapport, Livsmedelsverket, Sweden, 1994. http://www.livsmedelsverket.se/globalassets/livsmedel-innehall/naringsamnen/livsmedelsdatabas/Nutrient_losses_gains.pdf, (accessed November 2018).
54. A. L. Vásquez-Caicedo, S. Bell and B. Hartmann, Report on collection of rules on use of recipe calculation procedures including the use of yield and retention factors for imputing nutrient values for composite foods (D2.2.9), March 2008, European Food Information Resource Consortium (EuroFIR). http://www.eurofir.org/?page_id=2695, (accessed November 2018).
55. S. M. Church, *BNF Nutr. Bull.*, 2015, **40**, 40–44.
56. H. Reinivuo, Inventory of recipe calculation documentations of EuroFIR partners. An annex to the report of "Proposal for the harmonisation of recipe calculation procedures". WP 2.2 Composite Foods, EuroFIR, 2007. http://www.langual.org/Download/Recipe Calculation/D2%202%2012%202%202%204%20(3).pdf, (accessed November 2018).
57. (a) H. Reinivuo and K. Laitinen, Proposal for the harmonization of recipe calculation procedures. WP2.2 Composite Foods, EuroFIR, 2007; (b) H. Reinivuo, S. Bell and M. L. Ovaskainen, *J. Food Compos. Anal.*, 2009, **22**, 410–413.
58. M. Machackova, A. Giertlova and J. Porubska, *How to Calculate Nutrient Content of Foods: A Guideline for Food Business Operators*, EuroFIR, 2015. http://www.eurofir.org/wp-content/uploads/2015/12/EUROFIR-RECIPE-GUIDELINE_FINAL.pdf, (accessed November 2018).

59. European Commission, Nutrition and Health Claims. https://ec.europa.eu/food/safety/labelling_nutrition/claims_en, (accessed May 2019).
60. M. Machackova, A. Giertlova, J. Porubska, M. Roe, C. Ramos and P. Finglas, *Food Chem.*, 2018, **238**, 35–41.
61. Produkto maistingumo apskaičiavimas. http://vmvt.lt/maisto-sauga/verslui/zenklinimas-ir-reklama/maisto-produktu-zenklinimas/informacijos-apie-1, (accessed September 2019).
62. Ponuka výpočtu výživových hodnôt výrobkov pre prevádzkovateľov potravinárskych podnikov. https://vmvt.lt/maisto-sauga/verslui/zenklinimas-ir-reklama/maisto-produktu-zenklinimas/informacijos-apie-1, (accessed September 2019).
63. Labelling for Food Food Databanks at QI. https://fdnc.quadram.ac.uk/labelling/, (accessed November 2018).
64. M. Chylkova, *Prakticka aplikace pozadavku potravinoveho prava*, Potravinarska komora Ceske Republiky, 2014. https://web.vscht.cz/~mikovak/management/Aplikace%20potr,%20pr%C3%A1va%20PK%2014.pdf, (accessed November 2018).
65. Fevia, *La Législation Européeene Relative a L'étiquetage*, 2014. http://www.afsca.be/fr/extra/fic/_documents/2014_05_13_FIC_Voedingswaarde_etikettering.pdf, (accessed November 2018).
66. A. Oliveira, *Information to Consumers Regulation (EU) No 1169/2011*, European FooD-STA, 2017. https://www.food-sta.eu/system/files/event_attachments/Webinar%20-%20Information%20to%20Consumer%20-%20Regulation%20%28EU%29%20N%C2%BA%201169_2011.pdf, (accessed November 2018).
67. NASCENT, Nutritional labelling software and claims: service, training and innovation offering to SMEs and industry in Europe. http://www.eurofir.org/our-resources/current-projects/nascent/, (accessed November 2018).
68. Statni Zemedelska a potravinarska inspekce, *Program Skoleni k vypoctu vyzivových udaju*, 2016, Prague, Brno, Czech Republic.
69. N. Koen, R. Blaaw and E. Wentzel-Vijoen, *S. Afr. J. Clin. Nutr.*, 2016, **29**, 13–21.
70. C. Gonzalez-Vallejo, B. D. Lavins and K. A. Carter, *Appetite*, 2016, **105**, 71–84.
71. D. Nelson, D. Graham and L. Harnack, *J. Nutr. Educ. Behav.*, 2014, **46**, 589–594.
72. K. Cannoosamy, P. Pugo-Gunsam and R. Jeewon, *J. Nutr. Educ. Behav.*, 2014, **46**, 334–340.
73. New Rules for Nutrition Labelling of Prepacked Foods, *FSAI News*, 2014, **16**(4), 4–5.
74. I. Sukova, *Oznacovani Potravin – pruvodce pro spotrebitele*, Ministerstvo zemedelstvi, Prague, 2014. http://www.bezpecnostpotravin.cz/UserFiles/publikace/Oznacovani%20potravin%20-%20posledni%20verze%20web.pdf, (accessed November 2018).

75. The British Dietetic Association, Food Fact Sheet, 2015. https://www.bda.uk.com/foodfacts/labelling.pdf, (accessed November 2018).
76. L'étiquetage nutritionnel devient obligatoire pour tous les aliments préemballés. https://www.fevia.be/fr/presse/letiquetage-nutritionnel-devient-obligatoire-pour-tous-les-aliments-preemballes, (accessed November 2018).
77. C. Hawkes, in *Innovations in Food Labelling*, ed. J. Albert, FAO. http://www.fao.org/docrep/018/i0576e/i0576e.pdf, (accessed November 2018), 2000.
78. DEFRA, Impact Assessment on a Statutory Instrument implementing Regulation (EU) 1169/2011 on the provision of food information to consumers, 2012. https://www.gov.uk/government/uploads/system/uploads/attachment_data/file/82667/consult-fic-ia-20121107.pdf, (accessed November 2018).

CHAPTER 4

Health Claims Regulation: Opportunities and Challenges in Europe

JUDITH BUTTRISS

British Nutrition Foundation, New Derwent House, 69–73 Theobalds Road, London WC1X 8TA, UK
Email: J.Buttriss@nutrition.org.uk

4.1 Introduction

On-pack claims about nutrition or health properties were found to be present on 26% of foods in a survey undertaken in five European countries.[1] Of these, 64% were found to be nutrition claims, 26% health claims, 6% were health-related ingredient claims, and 6% were symbols of one sort or another. The prevalence varied by country and by food category. Using the composition of these foods as a guide, those carrying claims were judged to have marginally better nutritional profiles than those without claims,[2] concurring with similar studies in the UK and Australia. This serves to emphasise the potential importance of claims in directing shopper choice.

4.2 What Does the Law Require?

In Europe, the use of nutrition and health claims in relation to foods, beverages and dietary supplements has been governed by Regulation No 1924/2006[3] since 2007. Before this time, there were voluntary schemes in EU Member States (ME), including Finland, the Netherlands, Sweden and the

Food Chemistry, Function and Analysis No. 22
Health Claims and Food Labelling
Edited by Siân Astley

Published by the Royal Society of Chemistry, www.rsc.org

UK, but no formal regulation. Regulation No 1924/2006 covers all foods, drinks and dietary supplements sold in the EU, and it applies to all commercial communications, including the food label, advertorials and other promotional materials.

The aims of Regulation No 1924/2006 are to: ensure that claims used in Europe are scientifically valid; harmonise claims made across the EU; provide useful and reliable information to consumers; and encourage innovation in the food industry. Any claim made should be truthful and not attempt to mislead consumers. Also, claims should not call into question the safety or nutritional content of other foods or the adequacy of a balanced diet. Claims must apply to the product as consumed, prepared according to the manufacturer's instructions, and the effects described in the claim must be understandable to consumers. Claims about preventing, treating or curing a disease are not permitted, neither are claims on alcoholic beverages (more than 1.2% alcohol) (other than reference to low/reduced alcohol or energy) or claims that suggest health could be affected by not consuming the product, make reference to a rate or amount of weight loss or recommendations of individual doctors and health professionals.

A scientific assessment of evidence in support of health claims is carried out by the European Food Safety Authority (EFSA), an independent scientific body charged with providing scientific advice to the European Commission (EC). EFSA's scientific opinion is used by the EC, alongside considerations of consumer understanding, to decide whether to approve or reject the claim.

4.3 What Types of Claim Are Covered?

4.3.1 Nutrition Claims

Nutrition claims describe what a food or drink contains in reduced or increased amounts, or does not contain, *e.g.* sugar-free, low fat, high fibre, energy reduced, high in polyunsaturated fatty acids, source of vitamin C. To make the nutrition claim, conditions of use must be followed. Permitted nutrition claims and their conditions of use are listed in an Annex to Regulation No 1924/2006 and on the EC website.[4] Conditions of use are consistent with the Food Information to Consumers Regulation (1169/2011), which came into force in December 2014 and governs nutrition labelling.[5] Examples of conditions of use include:

- Sugar-free: contains no more than 0.5 g of sugars per 100 g or 100 mL.
- Low in fat: contains no more than 3 g of fat per 100 g or, for a liquid, no more than 1.5 g of fat per 100 mL, except semi-skimmed milk, which must contain no more than 1.8 g of fat per 100 mL.
- High in fibre: at least 6 g of fibre per 100 g or at least 3 g of fibre per 100 kcal.
    - Threshold for source of fibre is 3 g/100 g.
- Energy reduced: energy content has been reduced by at least 30%.

- High in PUFA (polyunsaturated fatty acids): at least 45% of the fatty acids present are polyunsaturated and together these provide more than 20% of the energy in the product.
- Source of vitamin C: contains at least 15% of the nutrient reference value (NRV) for vitamin C per 100 g and, for a drink, at least 7.5% of the NRV for vitamin C per 100 mL. To claim high in, foods or beverages must provide at least 30% or 15%, respectively.
  - Reference intake (RI) is often used instead of NRV and the same rules apply for other vitamins and minerals for which NRVs have been set but, if a nutrient is not on the list, a claim cannot be made.

4.4 Health Claims

A health claim is any statement about a relationship between food and health.[6] A list of health claims that are permitted or have been rejected has been published by the EC.[7] Scientific opinions from the EFSA Dietetic Products, Nutrition and Allergies (NDA) Panel on each of the claims are published in EFSA's official journal.[8] Regulation No 1924/2006 sets out the approval processes for the different categories of health claims, which are known as Article 13.1, Article 13.5, Article 14(a) and Article 14(b) health claims (Figure 4.1).

4.5 Article 13.1 Health Claims

Article 13.1 health claims are those supported by "generally accepted scientific evidence". However, there has been some confusion as to what this means exactly. Generally, the claims accepted onto this list are those supported by substantial scientific evidence that is already in the public domain and is likely to be found, for example, in a standard nutrition science text book. Examples of Article 13.1 health claims that have been approved include:

- Vitamin D contributes to the maintenance of normal bones
- Folate contributes to maternal tissue growth during pregnancy
- Protein contributes to the maintenance of muscle mass
- Iron contributes to normal oxygen transport in the body.

The word "contributes" often features, emphasising the fact that physiological functions referenced often require a number of nutritional and other inputs. Also, "normal" is used commonly in these permitted claims, yet research amongst consumers suggests this terminology is not well understood.

The evaluation procedure for Article 13.1 claims began once Regulation No 1924/2006 was introduced, with competent authorities from each MS compiling a list of potential claims and submitting these to EFSA over the period 2008–10. A list of supporting references was provided for each claim,

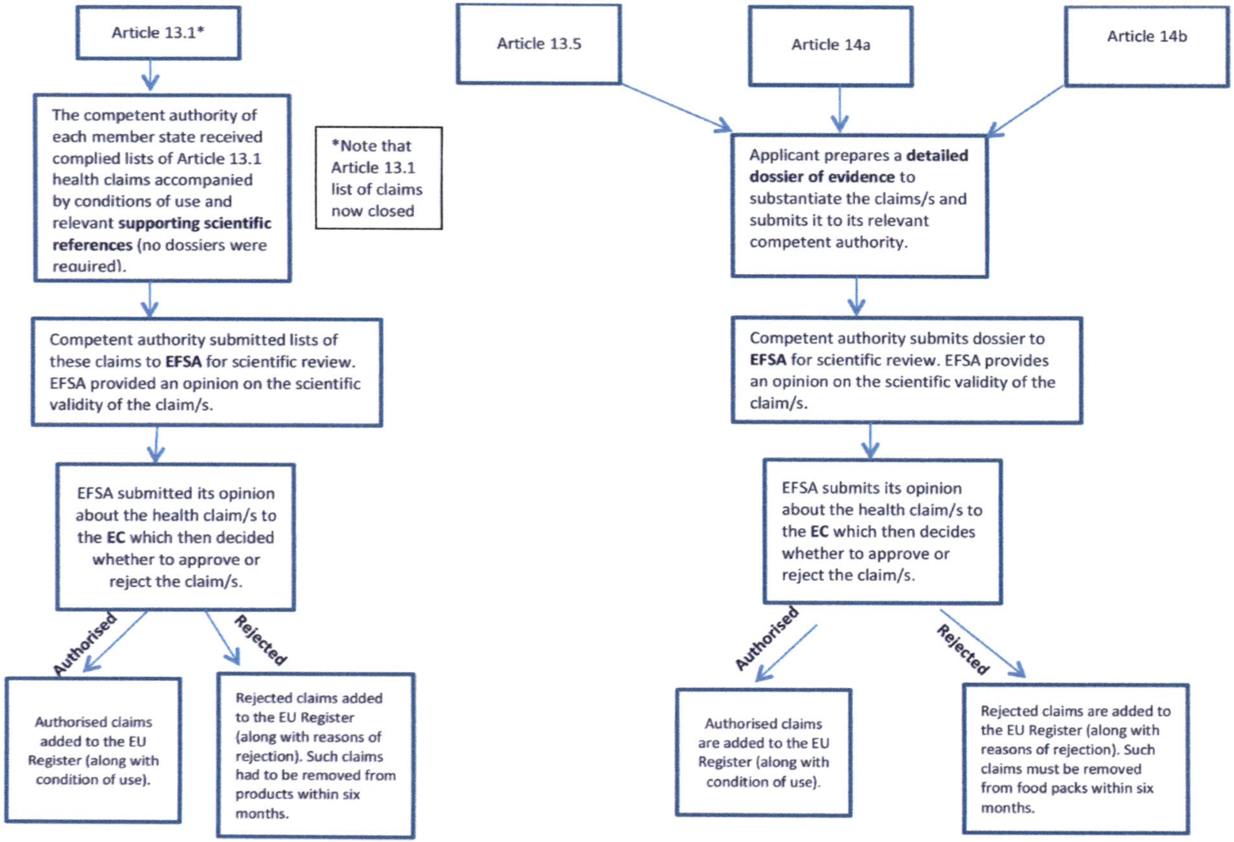

Figure 4.1 Representation of the submission and processing procedures for authorisation of Article 13.1, 13.5, 14a and 14b health claims.

but a full dossier of evidence was not required because these claims are intended to be supported by "generally accepted scientific evidence". Approximately 40 000 potential health claims were submitted. The list was whittled down to about 4000 claims once duplications and ineligible claims had been removed, which EFSA assessed on behalf of the EC. Then, in May 2012, the EC published lists of accepted and rejected Article 13.1 claims in the Community Register of Claims.[7] These claims officially entered into law on 14 June 2012 and food business operators were required to comply with the published list by December 2012. The Article 13.1 list of health claims is now, in effect, closed, although a large number of claims on botanical substances are currently on hold (see Chapter 2).

4.6 Article 13.5 Health Claims

Article 13.5 health claims are those based on more recently generated and, potentially, less established scientific evidence. Unlike Article 13.1 claims, applicants wishing to make an Article 13.5 claim must submit a detailed dossier. This is considered by EFSA, which provides the EC with an opinion on the strengths and weaknesses of the evidence, but it is the EC that makes the final decision to accept or reject the claim, considering consumer understanding and other factors. Article 13.5 permits applications for proprietary health claims, *i.e.* claims that are reliant for substantiation on data owned by a company or individual and not currently in the public domain. In this case, if approved, the health claim cannot be used by companies other than the applicant for five years following its publication on the health claims list (*e.g.* effect of a slowly digestible starch on postprandial blood glucose[9] or effect of cocoa flavanols on the elasticity of blood vessels[10]). A proprietary health claim can only be obtained as part of a successful health claim application if EFSA agrees that the claim could not have been established without proprietary data. If the same conclusion can be drawn from published data, the health claim is unlikely to be granted proprietary status.

4.7 Article 14 Health Claims

Article 14 health claims are those that relate to either reduction of disease risk (14a) or to children's health and development (14b). As with Article 13.5 claims, a detailed dossier of evidence must be submitted in order to apply for an Article 14 health claim. Although claims about disease risk reduction are permitted (14a), the claim must refer to the effect on a risk factor rather than the disease directly. For example, reduction of blood cholesterol, as a risk factor for heart disease, rather than reduced risk of heart disease. This distinguishes Regulation No 1924/2006 from rules that exist in some other countries, in particular the USA, and ensures a distinction between foods and medicines.

All of the words in an Article 14a health claim (*i.e.* both sentences) have to be included and, for some claims, there are specific conditions of use, detailed in the published record.[7] Examples of approved Article 14a health claims include:

- Plant sterols and plant stanol esters have been shown to lower blood cholesterol. High cholesterol is a risk factor in the development of coronary heart disease.
- Oat beta-glucan has been shown to lower blood cholesterol. High cholesterol is a risk factor in the development of coronary heart disease.

Health claims referring to the development and health of children are covered by Article 14b and should be substantiated with data obtained in studies conducted specifically in children. As with Article 13.1 claims, "normal" often features in the wording, which also is often quite detailed. Examples of approved Article 14b claims include:

- Calcium is important for normal growth and development in children.
- Protein is needed for normal growth and development of bone in children.
- Docosahexaenoic acid intake contributes to the normal visual development of infants up to 12 months of age.

4.8 Reference to General Non-specific Health Benefits

Article 10.3 of Regulation No 1924/2006 deals with references to general non-specific benefits of nutrients or foods for overall good health or health-related well-being. Such statements can only be made if accompanied by a specific authorised health claim, which must appear next to or following the general statement. For example, a product may claim: "Contains vitamin C. Vitamin C contributes to the normal function of the immune system", provided the food or beverage meets the conditions of use for "source of", which is at least 15% of the NRV for a food.

4.9 Nutrient Profiles

Regulation No 1924/2006 includes provision for a nutrient profiling system, which would restrict the use of both nutrition and health claims to products that meet all of the nutrient profiling criteria. For nutrition claims, identified as "any claim that states, suggests or implies that a food has a particular beneficial nutritional property due to the presence, absence, increased or reduced levels of energy or of a particular nutrient", and includes claims such as "low fat", Regulation No 1924/2006 states these would be permitted on products that fail to meet no more than one criterion, but the nutrient threshold that is exceeded will have to be stated on pack, *e.g.* "low fat, high in sodium". However, the nutrient profiling system, scheduled for implementation in 2009, has yet to emerge (see Chapter 2).

4.10 Regulation No 1924/2006 Evaluation: Process in Outline

In summary, for Article 13.5 and 14 health claims, a dossier to support the application must be prepared including the applicant's details, characterisation of the substance for which the claim is being made, a comprehensive review of the evidence for the substance's effect on health, and a proposal for the wording of the claim. This is considered by EFSA on behalf of the EC (see Figure 4.1).

Characterisation has emerged as a particularly important aspect of submissions, as only once the characterisation has been assessed as acceptable will scientific information supporting the claim be assessed; good science describing the health effects could, potentially, be undermined by weak characterisation of the active compound(s). Insufficient characterisation has proved to be the downfall of several applications, in particular applications concerning probiotic strains. Characterisation must include details of composition, physical and chemical characteristics, manufacturing process, stability and bioavailability.

Applications must contain all of the scientific data relevant to the claim, both in favour of the claim and against, in order to demonstrate the health claim is substantiated by the totality of the evidence. In practice, this means a comprehensive review of human studies addressing the specific relationship between the food/constituent and the effect claimed, performed in a systematic and transparent manner and considering the hierarchy of evidence shown in Table 4.1.[11]

The EFSA NDA Panel provides an opinion on the scientific validity of the health claim, using the hierarchy of evidence. If it is decided that a cause-and-effect relationship has been established, EFSA will consider whether the proposed wording reflects the scientific evidence and complies with

Table 4.1 Hierarchy of evidence considered by EFSA. Reproduced from ref. 11 (https://doi.org/10.2903/j.efsa.2017.4680) under the terms of a CC BY-ND 4.0 license (https://creativecommons.org/licenses/by-nd/4.0/v).

Human intervention studies (original research)	(a) Randomised controlled studies (b) Other randomised studies (non-controlled) (c) Controlled, non-randomised studies (d) Other intervention studies
Human observational studies (original research)	(a) Cohort studies (b) Case-control studies (c) Cross-sectional studies (d) Other observational studies (*e.g.* case reports)
Review publications	(a) Systematic reviews, pooled analyses and meta-analyses (b) Other review publications
Other	For example, human studies dealing with the mechanisms by which the food/constituent could be responsible for the claimed effect

criteria laid down in Regulation No 1924/2006, ensuring the wording is "truthful, clear, reliable and useful to the consumer in choosing a healthy diet". If this is not the case, the NDA Panel may propose alternative wording. Guidance stresses that the wording should be easily understood by the "average consumer" and, for reduction of disease risk claims (14b), the wording should refer to a specific risk factor for the disease and not the disease.

The EC makes the final decision on whether to approve or reject the health claim. All claims are published in the Community Register of Claims as either authorised or non-authorised. Once a claim is non-authorised, food business operators have six months to stop using the claim in question.

In January 2017, EFSA published updated scientific and technical guidance for applicants on how to prepare and present a health claim application.[11] The update took account of a public consultation as well as experience gained by EFSA in evaluating health claims and interactions with stakeholders. The update was also aligned with general scientific guidance for stakeholders on health claim applications,[12] which outlined the evaluation process and gave applicants a better understanding of the assessment criteria.[13] In particular, the wording was adapted for claims based on the essentiality of nutrients, recognising that randomised controlled trials are not applicable in such cases (see Table 4.2).

The new guidance also provided a standardised format for a well-structured application. It detailed the kinds of information and data applicants needed to submit in support of claims. The guidance applies to health claims for a type of food, a food category or food constituents. In addition to the required administrative and technical information, dossiers must still include proposed wording for the claim; information on the characterisation of the food or food constituent; information about the claimed beneficial physiological effect; and all pertinent scientific data (see Figure 4.2 and Table 4.1[11,13]). Most significantly, the new guidance emphasised the importance of:

- defining and characterising the food or food constituent;
- demonstrating that the claimed effect is based on the essentiality of a nutrient or is a defined beneficial effect for the target population and can be measured *in vivo* in humans;
- demonstrating essentiality, *i.e.* that it is required for normal human body functions (it has an essential mechanistic role in a metabolic function and/or has the ability to reverse clinical signs and symptoms of its deficiency), it cannot be synthesised by the body or cannot be synthesised in amounts adequate to maintain the normal body function that is the subject of the claim, it must be obtained from a dietary source OR (for a disease reduction claim) a cause-and-effect relationship is established by considering the strength, consistency, specificity, dose–response and biological plausibility of the relationship;

Table 4.2 Summary of the required information.

Required information	Notes
Proposed wording for the health claim and specific conditions of use	No notes
Information on the characterisation of the food or food constituent	This includes information to establish that the quantity of the food/constituent and pattern of consumption required to obtain the claimed effect could reasonably be achieved as part of a balanced diet.
Information to allow characterisation of the claimed effect	For function claims, the specific function in the body should be specified. For reduction of disease risk claims, both the risk factor and the disease should be identified. For both types of claim, a rationale should be provided to explain how the proposed changes are beneficial physiological effects for the target population, together with outcome variable(s) and methods of measurement which could be used to assess such changes *in vivo* in humans.
All pertinent scientific data	Includes published and unpublished data, and data in favour and not in favour, that forms the basis of the substantiation of the health claim. High-quality data is expected: the hierarchy of evidence used to assess the claim is shown in Table 4.1. For claims based on the essentiality of nutrients, data from studies in humans addressing the relationship are required. Data from animal or other model systems can't substitute for human data but can be used as supporting evidence (*e.g.* to demonstrate biological plausibility and provide mechanistic evidence). A comprehensive (systematic and transparent) review of pertinent published human studies is required, with the procedure used to identify the studies depicted.

- demonstrating the quantity of the food/constituent and patterns of consumption required to obtain the claimed effect can reasonably be achieved as part of a balanced diet.

4.11 Novel Foods Legislation

Submissions for new health claims may, in some instances, involve novel foods or novel ingredients, which must first be authorised under Novel Food Regulation (EU) 2015/2283.[14]

Novel foods legislation dates back over 20 years and, since then, technological developments and scientific advice have evolved considerably. On

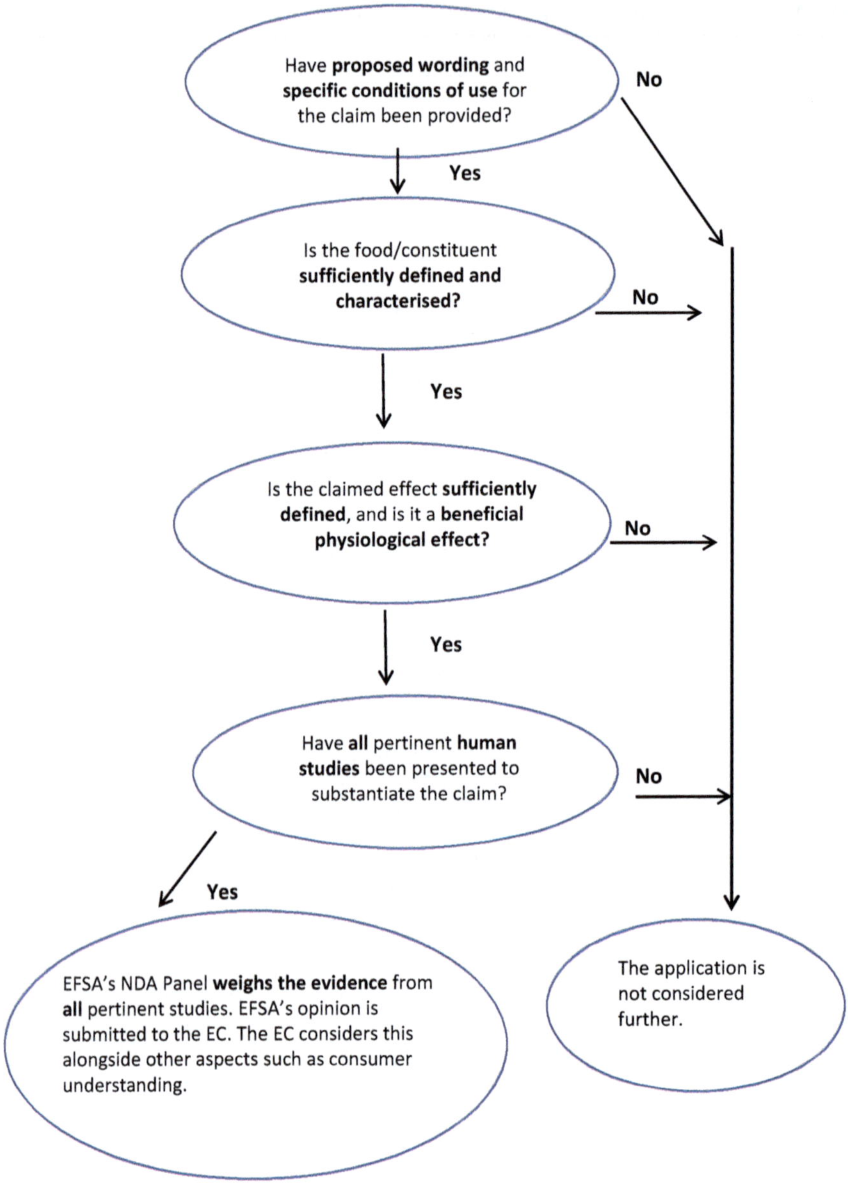

Figure 4.2 Key questions EFSA's NDA Panel considers in the scientific evaluation of health claims.
Reproduced from ref. 13 (https://doi.org/10.2903/j.efsa.2016.4367) under the terms of a CC BY-ND 4.0 license (https://creativecommons.org/licenses/by-nd/4.0/v).

review, the authorisation procedure was deemed to be too lengthy and, consequentially, was not actively encouraging innovation within the food and drink sector. In November 2015, an agreement was reached between the

European Parliament and the European Council on a new Novel Food Regulation 2015/2283.[15]

The newly adopted Regulation applied from 1 January 2018 (two years ahead of the date of entry into force) and replaces Novel Food Regulation No. 258/97.[16] In it, the definition of novel food has been updated and clarified, and explicitly includes whole insects and their parts, and engineered nanomaterials. The main objective of Novel Food Regulation 2015/2283 is to simplify the process, reducing the administrative burden and improving competitiveness and innovation, whilst ensuring food safety. Novel Food Regulation 2015/2283 also aims to simplify administrative procedures for public authorities including enforcement of the Regulation.

4.12 Consumer Understanding of Health Claims

Finding ways to help the general public understand the complex wording of claims has proved challenging.[17] However, the more consideration given to this matter during the compilation of health claim dossiers, the greater the likelihood that communication of the benefits will be straightforward and of value to the consumer. A number of EU MS have published guidelines on this and this topic has also been addressed in the most recent EFSA guidance.[13]

Regulation No 1924/2006 states that the beneficial effects expressed in a health claim must be understood by an average consumer, defining this as someone who is "reasonably well informed and reasonably observant and circumspect".[18] However, a definition of consumer understanding and an agreed method by which it can be quantified were not provided, and this has since proved problematic. This difficulty was addressed in the EC-funded CLYMBOL (2012–16), which considered how health claims are understood by consumers and the ways in which they influence purchase and consumption patterns in different countries, with the goal of delivering a method to assess the effects of health claims.[19]

An output from CLYMBOL was a methodology to explore whether individual claims are understood by the "average consumer", how understanding of claims and symbols can be improved, and whether claims and symbols trigger healthier choices.[20] A technique, now widely used in such research, is eye tracking, which enables information to be collected in real-world situations (*e.g.* supermarket aisles) rather than in a laboratory. This revealed the tiny window of milliseconds during which any packaging information – claim or nutrition label – has to attract the attention of shoppers in the context of all other factors that drive choice and purchasing habits, including price, taste and texture, familiarity and family preferences, convenience and pack size.

CLYMBOL demonstrated that consumer perceptions and understanding of a claim differ amongst EU MS,[19] meaning particular attention should be given to both the target group and geographical area when considering whether a claim is clearly communicated and easily understood. Research also suggests that consumers perceive a product to be of greater benefit and are more likely to pay attention if the claim relates to a health status they are familiar or associated with (*e.g.* those with a history of high blood cholesterol might pay more attention to a related claim[21]). Furthermore, the presence of (nutrition) claims might lead people to believe that foods carrying claims are healthier than similar foods without claims.[22]

The EC considers consumer understanding in its decision to authorise or reject a claim. Once the EC has approved a claim, food business operators are expected to use the authorised wording, although Regulation No 1924/2006 allows a limited amount of flexibility, with a few caveats. Minor changes in the wording used can affect greatly the way in which a claim is perceived. In 2012, EU MS agreed informally the following:[23]

- Adapted wording must have the same meaning as the authorised wording.
- 'Normal' must not be replaced or removed.
- Claimed effects must be attributed to the authorised nutrient, substance, food or food category and not the finished product.
- Particular care should be taken when making claims about supplements and it must be made clear which component of the supplement provides the claimed health effect.
- Any general, non-specific health claims on a product must be made in conjunction with an authorised Article 13 or Article 14 claim.
- Any brand names, trademarks or fancy names that denote general or non-specific health claims must be used in conjunction with an authorised Article 13 or Article 14 claim.
- Claims must not be adapted using phrases taken from an EFSA opinion.

Responsibility for monitoring the use of claims lies with individual EU MS, which has introduced variability in how communications are assessed and simplification of the communication is justified. Consequently, food products sold across the EU can be interpreted differently from one MS to another, leading to different ramifications and solutions, and thereby undermining the principle of consistent communication.[17]

4.13 Impact of the Regulation on Small Businesses

SMEs make up 99% of Europe's food and drink companies and account for 52% of the financial turnover in the European food industry.[24] Regulation No 1924/2006 aims to "respect fair competition and protect innovation in the area of foods", because business practices based on the use of unsubstantiated claims not only mislead consumers but also create conditions of unfair competition.[25] Yet SMEs often lack staff who are specialists in nutrition science and might not have in-house regulatory affairs expertise, meaning they struggle to access the expertise necessary to determine whether a claim application is feasible or desirable, and if further research to support the claim is needed.

The findings of a survey amongst SMEs, conducted by BACCHUS[17] and described in the BACCHUS Best Practice Guide,[26] found 90% were aware of Regulation No 1924/2006, but cited the financial cost of submitting an application dossier (*e.g.* research and the wider process) and the length of time as barriers. Just over half of those considering submission of a dossier (*ca.* 10) were planning to do this with proprietary data. The vast majority (75%) of respondents were aware of the Novel Food Regulation (EU) 2015/2283, with four respondents having first-hand experience of taking an ingredient through the authorisation process and 11 respondents currently selling products containing authorised novel foods or novel food ingredients.

Of course, it is paramount that the public is protected, and nutrition and health claims as well as novel foods and food ingredients are regulated appropriately. However, given the importance of SMEs to European economies, it is also important to ascertain whether innovation, viability and competitive advantage is affected adversely by policies introduced for the best reasons but perceived as complex, time-consuming and expensive and not understood by the average consumer. An EU-funded project, REDICLAIM (REduction of DIsease risk CLAIMs on food and drinks)[27] explored how Regulation No 1924/2006 has affected substantiation and use of reduction of disease risk claims (14b) on foods and drinks and concluded that the impact of this legislation depends on how it has been implemented and regulated in different EU MS. Effective regulation of health claims enables and enhances innovation whilst still being ethical. Gaps in the legislative framework and differences in how the legislation is applied not only create barriers against free movement of goods and fair competition but reduce interest in health-related research amongst food and beverage SMEs and larger food business operators.

4.14 Conclusions

Claims about nutritional and health properties of foods and beverages are regulated in Europe to protect the public, with detailed processes in place to ensure the scientific validity of potential claims. Regulation No 1924/2006 provides a level playing field for food business operators that wish to make the most of positive attributes of their food products but, over time, a number of difficulties with this legislation have emerged. Some of these have been addressed in the guidance published with the Regulation, which is updated from time to time, but others have proved to be more challenging. Regulation No 1924/2006 states that the wording of claims needs to be understood by the "average consumer" but this is not defined specifically and some of the physiological benefits associated with authorised claims are difficult to summarise simply. Another challenge for food business operators across EU MS is that monitoring of compliance is managed at the national level, and approaches vary amongst countries, inhibiting free movement of goods and fair competition as well as investment in health-related research: authorisation requires high-quality scientific evidence, which is time-consuming and costly, and may still be rejected by EFSA, the EC or national authorities. Constraints such as these have arguably limited the number of claims being submitted for scrutiny and might disadvantage disproportionately the smaller food business operators, which make up the majority of the European food and beverage sector.

Acknowledgements

The SME survey summarised in this paper was conducted under the auspices of a European Union 7th Framework Programme known as BACCHUS (FP7/2007-2013) under grant agreement 312090.

References

1. S. Hieke, N. Kuljanic and I. Pravst, *et al.*, Prevalence of nutrition and health-related claims on pre-packaged foods: a five country study in Europe, *Nutrients*, 2016, **8**, 1–16.
2. A. Kaur, *et al.*, The nutritional quality of foods carrying health-related claims in Germany, The Netherland, Spain, Slovenia and the United Kingdom, *Eur. J. Public Health*, 2016, **70**, 1388–1395.
3. Regulation (EC) No 1924/2006 of the European Parliament and of the council of 20 December 2006 on nutrition and health claims made on foods, http://data.europa.eu/eli/reg/2006/1924/2012-11-29.
4. European Commission, Nutrition Claims, https://ec.europa.eu/food/safety/labelling_nutrition/claims/nutrition_claims_en.
5. Regulation (EU) No 1169/2011 of the European Parliament and of the Council of 25 October 2011 on the provision of food information to consumers, http://data.europa.eu/eli/reg/2011/1169/oj.
6. European Commission, Health Claims, https://ec.europa.eu/food/safety/labelling_nutrition/claims/health_claims_en.
7. EU Register of nutrition and health claims made on foods, http://ec.europa.eu/nuhclaims.
8. EFSA Journal, Online ISSN 1831-4732, https://efsa.onlinelibrary.wiley.com/journal/18314732.
9. Scientific Opinion on the substantiation of a health claim related to "slowly digestible starch in starch-containing foods" and "reduction of post-prandial glycaemic responses" pursuant to Article 13(5) of Regulation (EC) No 1924/2006, https://www.efsa.europa.eu/en/efsajournal/pub/2292.

10. Scientific Opinion on the modification of the authorisation of a health claim related to cocoa flavanols and maintenance of normal endothelium-dependent vasodilation pursuant to Article 13(5) of Regulation (EC) No 1924/2006 following a request in accordance with Article 19 of Regulation (EC) No 1924/2006, https://www.efsa.europa.eu/en/efsajournal/pub/3654.
11. D. Turck, J.-L. Bresson and B. Burlingame, *et al.*, Scientific and technical guidance for the preparation and presentation of a health claim application (Revision 2), *EFSA J.*, 2017, **15**(1), 4680.
12. General scientific guidance for stakeholders on health claim applications, https://www.efsa.europa.eu/en/efsajournal/pub/4367.
13. European Food Standards Authority, General scientific guidance for stakeholders on health claims applications, *EFSA J.*, 2016, **14**(1), 4357.
14. European Commission, Legislation, https://ec.europa.eu/food/safety/novel_food/legislation_en [accessed September 2018].
15. Regulation (EU) 2015/2283 of the European Parliament and of the Council of 25 November 2015 on novel foods, http://data.europa.eu/eli/reg/2015/2283/oj [accessed September 2018].
16. Regulation (EC) No 258/97 of the European Parliament and of the Council of 27 January 1997 concerning novel foods and novel food ingredients, http://data.europa.eu/eli/reg/1997/258/oj [accessed September 2018].
17. J. L. Buttriss, Nutrition and health claims in practice, *Nutr. Bull.*, 2015, **40**, 211–222.
18. European Commission (EC) (2007) EC Regulation No 1924/2006on nutrition and health claims made on foods. The Official Journal of the European Union L12/18, http://eur-lex.europa.eu/LexUriServ/LexUriServ.do?uri=OJ:L:2007:012:0003:0018:EN:PDF [accessed August 2017].
19. S. Hieke, N. Kuljanic and J. M. Wills, *et al.*, The role of health-related claims and health-related symbols in consumer behaviour: Design and conceptual framework of the CLYMBOL project and initial results, *Nutr. Bull.*, 2015, **40**, 66–72.
20. CLYMBOL Outcomes, www.clymbol.eu/outcomes [accessed September 2018].
21. V. J. van Buul and F. J. P. H. Brouns, Nutrition and Health Claims as Marketing Tools, *Crit. Rev. Food Sci. Nutr.*, 2015, **55**, 1552–1560.
22. Z. Talati, S. Pettigrew and B. Neal, *et al.*, Consumers' responses to health claims in the context of other on-pack nutrition information: a systematic review, *Nutr. Rev.*, 2017, **75**(4), 260–273.
23. General Principles on Flexibility of Wording for Health Claims, https://www.gov.uk/government/uploads/system/uploads/attachment_data/file/217005/health-claims-flexibility-of-wording-principles-UK-19-Dec-2012.pdf [accessed September 2018].
24. Food Drink Europe, Figures on SME's, 2015, http://www.fooddrinkeurope.eu/industry-in-focus/topic/small-and-medium-sized-enterprises-smes/figures/ [accessed September 2018].

25. Parliamentary questions, answer given by Mr Dalli on behalf of the Commission, June 2012, European Commission. E-004428/2012, http://www.europarl.europa.eu/sides/getAllAnswers.do?reference=E-2012-004428&language=NL [accessed September 2018].
26. Introduction to the BACCHUS best practice guide for health claims http://www.eurofir.org/bacchus/resources/introduction-to-the-bacchus-best-practice-guide-for-health-claims/ [accessed September 2019].
27. Reduction of disease risk claims on food and drinks (REDICLAIM) Project ID 63036, https://cordis.europa.eu/project/rcn/109352_en.html [accessed September 2018].

CHAPTER 5

Use of Nutrition and Health Claims and Symbols on Prepacked Foods in Europe: From Consumer Exposure to Public Health Implications

IGOR PRAVST,*[a] ANITA KUŠAR,[a] KRISTA MIKLAVEC,[a] SOPHIE HIEKE,[b] MONIQUE RAATS[c] AND MIKE RAYNER[d]

[a] Nutrition Institute, Ljubljana, Slovenia; [b] European Food Information Council (EUFIC), Brussels, Belgium; [c] Food, Consumer Behaviour and Health Research Centre, University of Surrey, Surrey, UK; [d] British Heart Foundation Centre on Population Approaches for Non-Communicable Disease Prevention, Nuffield Department of Population Health, University of Oxford, Oxford, UK
*Email: igor.pravst@nutris.org

5.1 Introduction

Nutrition is an important factor influencing human health. As well as being a source of essential nutrients, foods can convey numerous other health benefits, which is the basis of functional foods that are usually labelled with nutrition and health claims. Such claims can be a powerful marketing tool, and can influence food choices.[1,2]

Food Chemistry, Function and Analysis No. 22
Health Claims and Food Labelling
Edited by Siân Astley

Published by the Royal Society of Chemistry, www.rsc.org

To ensure that claims do not mislead consumers, nutrition and health claims (NHC) are regulated in most developed countries. In the European Union (EU), regulations were harmonised in 2006 by Regulation (EC) No 1924/2006, which requires that claims are substantiated by generally accepted scientific data.

Regulation (EC) No 1924/2006 distinguishes different types of claims that can be used for labelling and promotion (including advertising). Nutrition claims are any statement stating, suggesting or implying that a food has particular beneficial (nutritional) properties due to the calorific value or composition with respect to the presence (or absence) of nutrients or other substances[3] (see Chapter 2). Only those claims listed in the Annex of Regulation (EC) No 1924/2006 are permitted, but they must also be in accordance with the conditions set for use. Any claim considered to have the same meaning for consumers as those in the Annex is subject to the same conditions of use. A claim that a food contains a nutrient or substance for which conditions are not specified in Regulation (EC) No 1924/2006, or any claim likely to have the same meaning for consumers, may only be made where the nutrient/substance is present in the final product in a significant quantity that will produce the effect claimed, as established by generally accepted scientific evidence. The substance must also be in a bioavailable form.[3,4]

Health claims refer to a relationship between a food (category/constituent) and health. Only authorised health claims can be used, with the exception of some health claims that are still in the process of scientific evaluation[5] (*i.e.* botanicals; see Chapter 2). The quantity of the food/constituent and pattern of consumption required to obtain the claimed beneficial effect must be included in the labelling and reasonably expected to be consumed in the context of a varied and balanced diet. The claim must be based on generally accepted scientific data and well understood by the average consumer.

The use of general (non-specific) health claims is also possible, if they are accompanied by a related authorised specific claim.[3] All authorised health claims are listed in a public Community Register of NHC,[3] which includes the wordings of claims and conditions of use, along with any restrictions. Regulation (EC) No 1924/2006 distinguishes three categories of health claims (Table 5.1). General function claims (Articles 13.1 and 13.5) describe the role of a food in body functions, including the sense of hunger or satiety, and do not refer to children; reduction of disease risk claims (Article 14a) state that consumption of a food/constituent significantly reduces a risk factor for development of a human disease. Regulation (EC) No 1924/2006 also mentions claims referring to children's development and health (Article 14b), for which no further definition is given[3,4] (see Chapter 4).

Regulation (EC) No 1924/2006 has affected the market strongly, as many health claims prior to the Regulation were rejected, meaning the labelling of many prepacked foods has changed subsequently.[6]

Table 5.1 Types of health claims according to EC regulations (data from ref. 3).

General function health claims	Based on generally accepted scientific evidence	Claims not referring to children and describing the role of a nutrient or other substance in the growth, development and the functions of the body, psychological and behavioural functions, slimming or weight control, *etc.*
	Based on new evidence	General function health claims based on newly developed scientific evidence
Claims referring to children's development and health		No further definition provided
Reduction of disease risk claims		Claims that state, suggest or imply that the consumption of a food category, a food or one of its constituents significantly reduces a risk factor in the development of a human disease
General (non-specific) health claims		References to general, non-specific benefits of the nutrient or food for overall good health or health-related well-being (may only be made if accompanied by authorised specific health claim)

5.1.1 Prevalence of Nutrition and Health-related Claims on Prepacked Foods

NHC might help consumers identify foods that are healthier. A number of studies have been conducted around the world to investigate the prevalence of claims on prepacked foods, particularly in the USA,[7,8] Canada,[9,10] Australia[11–14] and New Zealand.[15] In the EU, such studies were conducted in the UK,[16] Ireland[17] and Slovenia.[18] In 2013, as part of the EU-funded project CLYMBOL,[19] a study was conducted in five EU Member States (MS) to investigate the food categories in which NHC were most common, which types of claims were used, and to what nutrients and health relationships they referred.[20]

The frequency of different claims varied amongst food categories. In the early 2000s, research in the USA reported a much higher prevalence of nutrition claims (50%) in comparison with health claims (5%).[21] Research done a few years later found there were fewer nutrition claims (40%) and more health claims (9%).[8] In Canada, the proportion of food packaging with nutrition claims was 46%; health claims were not included in the research.[10]

In China, 12% of products included nutrition claims.[22] Ref. 12 showed that in Australia and New Zealand, 15% of products included health claims, 5% of which did not comply with legislation.

The CLYMBOL study included five European MS (Germany, the Netherlands, Slovenia, Spain and the UK) and found 26% of products were labelled with claims, of which approximately one-third were health claims whilst the other two-thirds were nutrition claims (Table 5.2). Differences amongst selected EU MS were only observed for health claims.[20] General health claims appeared on 7% of foods whilst only 5% of foods carried a nutrient and other function claim.

The highest penetration of nutrition claims was observed in the UK (29%) followed by Spain (23%), Slovenia (19%), the Netherlands (17%) and Germany (16%). Less variation across the EU MS was observed for health claims; the highest prevalence was observed in the Netherlands (14%) and the lowest in Spain (7%).[20]

Reduction of disease risk claims and children's development and health claims occurred on 0.6% and 0.7% of foods, respectively. The prevalence of reduction of disease risk claims was highest in the UK (1%) and lowest in the Netherlands (0.2%), whilst children's development and health claims ranged from 0% (in the Netherlands) to 1.3% (in Germany).[20]

More than one-third of nutrition claims referred to vitamins (22%; most commonly vitamin C) and/or minerals (13%; most commonly calcium). About a quarter of nutrition claims referred to fat content, 12% to sugar content and 9% to fibre.

Looking at health claims, 36% referred to an unspecified constituent (*e.g.* "Complete nutrition for optimal growth"), 21% to the whole food, and 16% to vitamins or minerals. Nutrient and other function claims most commonly referred to function of the digestive, metabolic and endocrine systems (40%; 57% of which referred to general metabolic functions and/or weight management functions), cardiovascular, haematological,

Table 5.2 Prevalence of nutrition claims, health claims and symbols in five European countries (CLYMBOL study) ($N = 2034$). Adapted from ref. 20 (https://doi.org/10.3390/nu8030137) under a CC-BY 4.0 licence (https://creativecommons.org/licenses/by/4.0/).

Claim type	% of foods with claim (95% CIs)
Nutrition claim	20.8 (19.0–22.5)
Nutrient content claim	*19.6 (17.8–21.3)*
Nutrient comparative claims	*2.4 (1.7–3.1)*
Health-related ingredient claim	3.5 (2.7–4.3)
Health claim	10.9 (9.6–12.3)
General health claim	*6.7 (5.6–7.8)*
Nutrient and other function claim	*5.2 (4.2–6.2)*
Reduction of disease risk claim	*0.6 (0.2–0.09)*
Children's development and health claims	*0.7 (0.4–1.1)*
Any type of claim (NHC)	26.0 (24.0–27.9)

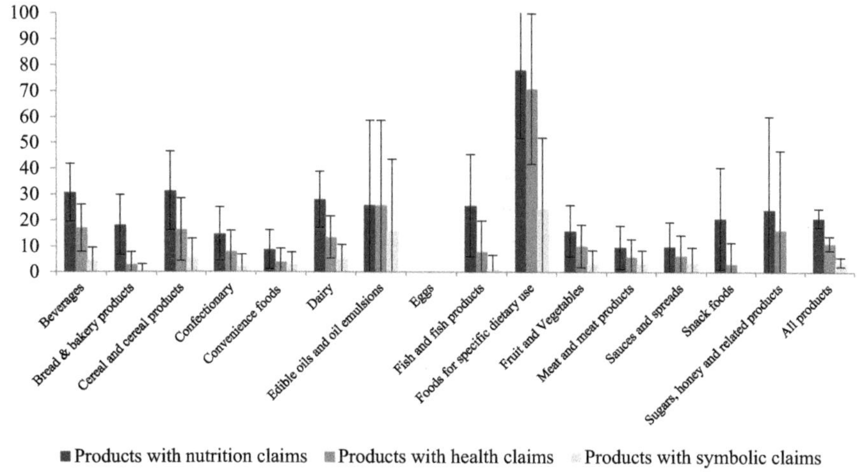

Figure 5.1 Prevalence of NHC (text or symbols) in the CLYMBOL study, by food category. (*Note*: % of foods with claim, including 95% confidence intervals.) Reproduced from ref. 20 (https://doi.org/10.3390/nu8030137) under a CC-BY 4.0 licence (https://creativecommons.org/licenses/by/4.0/).

immunological and respiratory systems (15%), mental function (11%) and neuro-musculoskeletal and movement-related functions (10%).[20]

The prevalence of NHC across food categories was also investigated (Figure 5.1).[20] The highest proportion of nutrition, health and symbolic claims (78%, 71% and 24%, respectively) was found in "foods for specific dietary uses", which included mostly baby foods. A high prevalence of nutrition claims (above 20%) was also found amongst "cereal and cereal products", "beverages" and "dairy products". Health claims were also found on "edible oils and oil emulsions", "fish and fish products" (both 26%), "beverages" (17%), "cereal and cereal products", "sugars, honey and related products" (both 16%) and "dairy products" (13%). The remaining food categories had less than 10% of foods carrying a health claim.

5.1.2 Nutritional Quality of Foods Labelled with Claims

To avoid a situation where claims could mask the overall nutritional status of a food product and mislead consumers when making food choices in the context of a balanced diet, Regulation (EC) No 1924/2006 introduced the concept of nutrient profiles, which should have been established by January 2009. However, as stated previously (see Chapters 2 and 4), this part of the legislation has not yet been implemented,[23] even though the scientific criteria were prepared in 2008.[24] This part of Regulation (EC) No 1924/2006 is still being evaluated as part of the European Commission's Regulatory Fitness and Performance program (REFIT)[25] and, therefore, it is not known whether nutrition profiles will ever be implemented or in what form.

In the absence of nutritional profiles, CLYMBOL also investigated the nutritional quality of foods carrying health claims.[26] Information from nutrient declarations (food labels) was assessed for mean levels of energy and nutrient content, using the Food Standards Australia New Zealand's Nutrient Profiling Scoring Criterion (FSANZ NPSC). Results showed that foods carrying health claims had marginally better nutritional quality than those without claims. Foods labelled with health claims had on average lower levels of energy (−29.3 kcal 100 g^{-1}), protein (−1.2 g), total sugars (−3.1 g), saturated fat (−2.4 g) and sodium (−842 mg), and higher levels of fibre (−0.8 g). A similar pattern was observed for foods labelled with nutrition claims.[26] Across the whole sample, 43% of foods passed the FSANZ NPSC, with foods carrying health claims more likely to pass (70%) than foods carrying nutrition claims (61%) or foods that did not carry any type of claim (36%).[26]

5.1.3 Ensuring Informed Consumer Choices

Research has shown different influences of food labelling information on consumers. Some results showed that the probability of product purchase increased when a health claim was present[27,28] whilst others suggested this is not the case.[7,21,29] There are some indications that consumers are unwilling to accept claims with which they are unfamiliar.[30,31] Additionally, incomprehensible language or concepts are confusing for consumers, and can negatively influence choice of the product.[32,33]

Labelling of prepacked foods is strictly regulated to support consumers' informed choices. Since 2016, all prepacked processed foods in the EU – with a few exceptions – must be labelled with nutrition declaration (Regulation (EU) No 1169/2011).[34] However, for foods labelled with nutrition or health claims, this requirement has been in place since 2007 (Regulation (EC) No 1924/2006);[3] and goes much further, requiring a nutrition declaration and the content of the nutrient/substance to which the claim refers. Products with health claims must include:

- A statement indicating the importance of a varied and balanced diet and a healthy lifestyle.
- Quantity and pattern of consumption required to achieve the claimed beneficial effect.
- Where appropriate, a statement addressed to those who should avoid using the product.
- An appropriate warning for products that are likely to present a health risk, if consumed in excess.

In case of reduction of disease risk claims, the labelling must also bear a statement indicating that the disease to which the claim refers has multiple risk factors and altering one of these may or may not have a beneficial effect.[3]

Regulation (EC) No 1924/2006 allows the use of easy, attractive statements that refer to the general, non-specific benefits of a product for overall good health or health-related well-being, without an authorisation procedure (Table 5.1). While the use of such statements could be helpful for consumers, as they should convey more consumer-friendly messages, such statements can easily be misunderstood, leading people to imagine different health benefits.[35] For this reason, such statements must be accompanied by authorised health claims next to or following the general statement. To avoid misleading consumers, manufacturers have to demonstrate the link between the general, non-specific benefits of the food and the specific, accompanying and authorised health claim.[35]

In the EU, rules related to NHC are harmonised at the community level, while each MS needs to ensure their enforcement. The rules were set to ensure food labels provide the necessary information that enables consumers to make informed choices. However, to our knowledge, there are no studies investigating this in real life. Instead, the studies have focussed primarily on the frequency and types of claims used, not on the other information provided on the labels of foods that carry NHC.

The objective of the CLYMBOL study was, therefore, to investigate the extent of information provided to consumers in a sample of prepacked foods labelled with nutrition and/or health claims.[20]

5.2 Methods and Materials

Altogether 2034 prepacked foods were sampled in five EU MS: Germany, the Netherlands, Slovenia, Spain, and the UK (approximately 400 foods per country were purchased using a randomised sampling approach). Of these, 466 products carried nutrition and/or health claims. Sampling took place during 2013 in three different outlets in each EU MS: a large supermarket, a discount store and a local (neighbourhood) shop. Food was defined as a single item available for sale in the selected outlet. This definition meant that the same food in different sizes could be included in the survey, on the basis that the packaging of these products might carry different claims. However, within-country duplicates were removed from the database. Non-food items, food supplements, alcoholic drinks and unpackaged foods were excluded. Detailed information about food sampling and data collection is reported elsewhere.[20]

Labelling data were recorded from packaging and coded. Worded claims were translated into English, while pictorial claims were described briefly. Other labelling and food composition data were also collected (*e.g.* whether packaging had front-of-pack nutrition labelling).[20] All products labelled with claims were examined for all the information that need to be provided, according to Regulation (EC) No 1924/2006 (*e.g.* nutrition declaration labelling, nutrient content, supporting statements where needed).

5.3 Results and Discussion

5.3.1 Labelling with Nutrition Declaration

The majority of the foods were labelled with a nutrition declaration (83%) (Table 5.3) although, at the time of sampling, this was only mandatory for foods labelled with claims. There were some differences amongst the EU MS; Slovenia has the lowest proportion of food products labelled with nutrition declarations (72%).

In the case of foods labelled with nutrition claims, where nutrition declarations are mandatory, such information was found on 96% of products. The lowest percentage of foods labelled with nutrition declarations was 89% (in the UK). In the case of products labelled with health claims, 81% provided nutrition declarations; this percentage was lowest in the Netherlands (71%), Slovenia (75%) and the UK (80%) and highest in Spain and Germany (90% for both countries). Almost all products that carried health symbols were also labelled with nutrition declarations (97%).

Considering the whole sample, the results were comparable to the FLABEL study conducted in 2009 across the EU27 plus Turkey.[36] In that study, more than 37 000 products were audited in a total of 84 retail outlets, and 85% of products were labelled with nutrition declarations (lowest was Slovenia with 70%, highest was Ireland with 97%). However, FLABEL included only a few categories of prepacked foods, whereas above average penetration of nutrition declarations was observed in our study, indicating that use of nutrition declarations has increased considerably, at least in countries where penetration rates were low in the past.

Nutrition declaration has been mandatory on prepacked foods for quite some time in many countries including the USA, Canada, Australia and New Zealand.[10,21,37] In comparison, nutrition declarations occur on about a quarter of products in China, where such labelling was not mandatory.[22]

Table 5.3 The penetration of nutrition declarations on complete samples of foods, and foods labelled with nutrition claims, health claims and symbols.

		% of foods labelled with nutrition declaration			
	Full sample (*N*=)	All foods (N = 2034)	Food labelled with nutrition claims (*N* = 389)	Food labelled with specific health claims (*N* = 128)	Food labelled with health symbols (*N* = 66)
Germany	399	81	100	*100*	/
Netherlands	416	86	97	*100*	96
Slovenia	416	72	96	*91*	100
Spain	405	88	99	*92*	100
UK	398	87	89	*100*	100
Total	**2034**	**83**	**96**	***97***	**97**

5.3.2 Labelling of Foods Carrying Nutrition Claims

In line with Regulations (EC) No 1924/2006 and No 1169/2011, nutrition labelling of products with claims is mandatory.[3] Such declarations can consist of: energy*, fat* (of which, saturates*, monounsaturates, polyunsaturates), carbohydrate* (of which sugars*, polyols, starch), fibre, protein*, salt*, and vitamins and minerals [*mandatory fields].[34] Where a nutrition claim is made for other constituents, the amounts of those substances cannot be part of the nutrition labelling. However, in such cases, the content needs to be stated in the same field of vision as the nutrition declaration and the units used to express amounts should be appropriate for the substances concerned.[3]

Altogether, about one-fifth of food products in the CLYMBOL sample were labelled with nutrition claims,[20] most (96%) of which include a nutrition declaration. The content of the nutrient/substance to which the nutrition claim referred was found on 90% of products (Table 5.4). The lowest ratios were observed in the UK sample, but this is because the UK market has a number of products with claims about non-nutrients.

In the EU, only authorised nutrition claims (or claims likely to have the same meaning for the consumer) can be used. When such claims are used, products need to meet conditions of use, as stated in the Annex of Regulation (EC) No 1924/2006. Three-quarters of the products in the CLYMBOL sample complied with these requirements (Table 5.4). Where this was not the case, the most common reasons were that products did not meet defined conditions for a certain claim (*i.e.* lower content of omega-3 than required in the Regulation) (7%) or it was not possible to determine whether conditions could be met (8%) (*e.g.* amounts of constituents were not provided) or, in a few examples, non-authorised nutrition claims were used (*e.g.* "free from cholesterol") (5%).

5.3.3 Labelling of Foods Carrying Health Claims

Considering that different regulatory conditions apply for use of specific and non-specific health claims, the CLYMBOL study explored these two

Table 5.4 Results related to the foods labelled with nutrition claims.

Country	Labelled nutrition declaration (%)	Labelled content of the food constituent (%)	Authorised and complying with conditions of use (%)
Germany	100	95	84
Netherlands	97	96	81
Slovenia	96	90	72
Spain	99	89	72
UK	89	85	71
Total	96	90	75

categories separately. In practice, non-specific health claims can be found on food labels as either:

- Article 3: Trademark (TM), brand name (BN) or fancy name (FN), which might be construed as a health claim.
- Article 10.(4): Reference to general, non-specific benefits of the nutrient or food for overall good health or health-related well-being.

About 17% of products labelled with health claims were categorised as a trademark (TM), brand name (BN) or fancy name (FN), which could be understood as a claim. In such cases, food labels need to be labelled with a specific claim and should meet all other requirements. However, classifying TM, BN and FN as health claims is subjective. Also, Regulation (EC) No 1924/2006 provided transitional measures, allowing products to bear TMs or BNs that existed before 2005 but do not comply with the Regulation. Considering this, such health claims (Art. 3) were excluded from detailed analyses but, whilst only 19% of such products were labelled with a specific claim, most (78%) carried nutrition claims.

About two-thirds (63%) of products were labelled with Article 10(4) general (non-specific) health claim. Of these, 13% did not have a nutrition declaration, and only 33% were labelled with a specific health claim (Table 5.5). Both these elements are mandatory. Furthermore, Regulation (EC) No 1924/2006 requires that a specific authorised health claim must be made next to or following general statements.[35] In our sample, specific health claims were in the same field of vision as general health claims in only 26% of products. General health claims were most frequent on Slovenian products (36%) but very rarely accompanied by a specific health claim (11%). More consistent labelling of general health claims was observed in Spain and the UK, where specific health claims were found on about half of the products labelled with general claims.

The CLYMBOL study was very strict about the interpretation of Article 10.(4) general (non-specific) health claims. Any reference to health or well-being was counted as a general (non-specific) health claim (*e.g.* cholesterol

Table 5.5 Results related to the foods labelled with general (non-specific) health claims.

Country	Labelled nutrition declaration (%)	Labelled specific health claim (%)	Labelled specific health claim in same field of vision (%)	Labelled also with a nutrition claim (%)
Germany	95	33	29	62
Netherlands	100	40	30	80
Slovenia	69	11	17	44
Spain	100	56	11	89
UK	96	54	42	96
Total	87	33	26	68

lowering, antioxidant, weight management, healthy, teeth-friendly, helps you to perform better, for life power, vitality for body and mind, *etc.*). It is inevitable that a label will not comply with Regulation (EC) No 1924/2006 if food manufacturers do not understand such statements to be a health claim. However, ultimately, interpretation is a matter for individual EU MS food authorities. EU MS have taken different approaches in implementing Regulation (EC) No 1924/2006 at national levels as well as the responsibilities of food authorities and control of the food supply,[38] which have been investigated further by REDICLAIM (Reduction of disease risk claims on food and drinks[39] – see also Chapter 4).

Specific health claims were most frequent in the UK, which was also the sample with highest penetration of other information specified by Regulation (EC) No 1924/2006 (*e.g.* nutrition declaration, content of the nutrient/substance, required mandatory statements). Only about half of the products with a specific health claim (46%, Table 5.6) were labelled with an authorised claim and compliant with the conditions of use. The most common issue observed in non-compliant labelling was use of unauthorised claims (18%), claims not submitted for evaluation (19%), unauthorised wording (10%) or reference to botanicals (7%).

Where reduction of disease risk claims is used, the label should also provide additional mandatory statements. In the CLYMBOL sample, only 0.6% of foods carried reduction of disease risk claims[20] (Table 5.2), but almost half of these (42%) lacked a statement indicating the disease has multiple risk factors and altering one of these risk factors may or may not have a beneficial effect overall.

5.4 Conclusions

The pan-European CLYMBOL study found that 21% of prepacked foods carry nutrition claims while health claims were found on 11% of foods, mostly in the form of general (non-specific) health claims (7%). Foods carrying claims had only marginally better nutritional quality than those that did not. Products labelled with health claims had, on average, lower levels of energy, protein, total sugars, saturated fat and sodium, and higher levels of fibre. A similar pattern was observed for foods labelled with nutrition claims. A little less than half of the foods sampled passed the FSANZ NPSC, but foods with health claims were more likely to pass (70%) than foods with nutrition claims (61%) or foods without any type of claim (36%).

CLYMBOL found that foods labelled with nutrition and/or (specific) health claims were labelled for the most part with mandatory nutrition declaration, the majority of which complied with Regulation (EC) No 1924/2006. A quarter of the foods with nutrition claims did not include the content of the nutrient/substance referred to in the claim or products did not meet defined conditions for an authorised claim while, in some cases, unauthorised claims were used. In cases of products labelled with general (non-specific) health claims, the occurrence of nutrition declaration was

Table 5.6 Results related to the foods labelled with specific health claims.

			Labelled with regulatory statements (%)				
Country	Labelled nutrition declaration (%)	Labelled content of the food constituent (%)	Healthy lifestyle statement[a]	Usage recommendation[b]	Population that should avoid using food[c]	Warnings[d]	Authorised and complying with conditions of use (%)
Germany	100	70	70	90	0	10	30
Netherlands	100	86	57	86	0	0	86
Slovenia	91	91	73	91	0	9	36
Spain	92	77	62	54	15	23	31
UK	100	92	83	83	13	13	46
Total	97	83	69	81	6	11	46

[a]A statement indicating the importance of a varied and balanced diet and a healthy lifestyle.
[b]The quantity of the food and pattern of consumption required to obtain the claimed beneficial effect.
[c]A statement addressed to persons who should avoid using the food – needed where appropriate.
[d]Warning for products that are likely to present a health risk if consumed to excess – needed where appropriate.

lower, possibly because some general claims were not recognised as health claims by the manufacturer. This presented a major issue for meeting the requirement that general (non-specific) health claims must be accompanied by a specific authorised health claim. This criterion was met in only about half of the products.

Development of food labels is an important step in delivery of prepacked foods that needs more attention from food manufacturers. CLYMBOL recommended that food business operators should ensure that NHC are not used on foods with unfavourable nutrition composition, which should be easier with the advent of Regulation (EC) No 1169/2011, *i.e.* mandatory nutrition declarations. Good practice was shown by some producers through the introduction of internal nutrition profiles to limit use of claims to healthier foods. Additionally, statements on non-specific benefits of a product for overall good health or health-related well-being need to be used with care. Such statements can be helpful for consumers, as they convey more consumer-friendly messages than many of the EFSA-authorised NHC, but these can also be misunderstood, misleading consumers about potential health benefits. Food business operators should be aware that such statements, as well as trademarks and brand names that might be construed as health claims, should be accompanied by authorised health claims.

Acknowledgements

The authors acknowledge the 7th EU Framework Programme, Small Collaborative Project CLYMBOL (Contract No. 311963) has been the major source of information for this article. The content of the article reflects only the views of the authors; the European Commission is not liable for any use that may be made of the information contained in this article. The study was also supported by the Slovenian Research Agency and the Ministry of Health of the Republic of Slovenia (V3-1501; L3-9290; P3-0395: Nutrition and Public Health). The funding organisations had no role in the design, analysis or writing of this article. The help of the following project partners in contributing to the development of the research protocol as well as the actual data collection and extraction is gratefully acknowledged: Kerry A. Brown, Bernadette M. Egan and Matthew Peacock (University of Surrey), Andrea Gröppel-Klein, Katja Pfeifer and Stephanie Leick (Saarland University), Živa Lavriša (University of Ljubljana/Nutrition Institute), Klaus G. Grunert, Violeta Stancu and Stewart Palmer (Aarhus University), Asha Kaur, Peter Scarborough and Anja Mizdrak (University of Oxford) and Anita Laser-Reuterswärd (Swedish National Food Agency), Nera Kuljanic (European Food Information Council, Brussels) and Azucena Gracia (Agrifood Research and Technology Centre of Aragon, Zaragoza, Spain).

References

1. A. Kaur, P. Scarborough and M. Rayner, *Int. J. Behav. Nutr. Phys. Act.*, 2017, **14**, 17.

2. N. Loebnitz, G. Schuitema and K. G. Grunert, *Psychol. Market*, 2015, **32**, 408.
3. (a) EU, Regulation (EC) No 1924/2006, Official Journal of the European Union, 30.12.2006, L404/9; (b) EU Register of nutrition and health claims made on food, http://ec.europa.eu/food/safety/labelling_nutrition/claims/register/public/?event=register.home.
4. I. Pravst, in *Dietary Supplements*, ed. K. Berginc and S. Kreft, Woodhead Publishing, Cambridge, 2015, pp. 3–24.
5. A. Kusar and I. Pravst, *Agro Food Ind. Hi-Tech*, 2014, **25**, 33.
6. A. Cavaliere, E. C. Ricci and A. Banterle, *Food Qual. Prefer.*, 2015, **41**, 44.
7. S. J. Brecher, M. M. Bender, V. L. Wilkening, N. M. McCabe and E. M. Anderson, *J. Am. Diet. Assoc.*, 2000, **100**, 1057.
8. S. E. Colby, L. Johnson, A. Scheett and B. Hoverson, *J. Nutr. Educ. Behav.*, 2010, **42**, 92.
9. J. E. Sacco, D. Sumanac and V. Tarasuk, *J. Nutr. Educ. Behav.*, 2013, **45**, 518.
10. A. Schermel, T. E. Emrich, J. Arcand, C. L. Wong and M. R. L'Abbe, *Appl. Physiol. Nutr. Metab.*, 2013, **38**, 666.
11. P. Williams, H. Yeatman, S. Zakrzewski, B. Aboozaid, S. Henshaw, K. Ingram, A. Rankine, S. Walcott and F. Ghani, *Asia Pac. J. Clin. Nutr.*, 2003, **12**, 138.
12. P. Williams, H. Yeatman, L. Ridges, A. Houston, J. Rafferty, A. Ridges, L. Roesler, M. Sobierajski and B. Spratt, *Asia Pac. J. Clin. Nutr.*, 2006, **15**, 10.
13. K. Z. Walker, J. L. Woods, C. A. Rickard and C. K. Wong, *Public Health Nutr.*, 2007, **11**, 1046.
14. C. Hughes, L. Wellard, J. Lin, K. L. Suen and K. Chapman, *Public Health Nutr.*, 2013, **16**, 2154.
15. A. Devi, H. Eyles, M. Rayner, C. Ni Mhurchu, B. Swinburn, E. Lonsdale-Cooper and S. Vandevijvere, *Appetite*, 2014, **81**, 253.
16. A. Kaur, P. Scarborough, A. Matthews, S. Payne, A. Mizdrak and M. Rayner, *Public Health Nutr.*, 2016, **19**, 588.
17. F. Lalor, J. Kennedy, M. A. Flynn and P. G. Wall, *Public Health Nutr.*, 2010, **13**, 704.
18. I. Pravst and A. Kušar, *Nutrients*, 2015, **7**, 9353.
19. (a) S. Hieke, N. Kuljanic, J. M. Wills, I. Pravst, A. Kaur, M. M. Raats, H. C. van Trijp, W. Verbeke and K. G. Grunert, *Nutr. Bull.*, 2015, **40**, 66; (b) CLYMBOL Project ID: 311963, Role of health-related claims and symbols in consumer behaviour, https://cordis.europa.eu/project/rcn/105145_en.html [accessed September 2019].
20. S. Hieke, N. Kuljanic, I. Pravst, K. Miklavec, A. Kaur, K. A. Brown, B. M. Egan, K. Pfeifer, A. Gracia and M. Rayner, *Nutrients*, 2016, **8**, 137.
21. L. Legault, M. B. Brandt, N. McCabe, C. Adler, A. M. Brown and S. Brecher, *J. Am. Diet. Assoc.*, 2004, **104**, 952.
22. J. Lv, Y. Chen, S. Wang, Q. Liu, Y. Ren, S. Karrar and L. Li, *Public Health Rep.*, 2011, **126**, 116.

23. F. P. Cappuccio and I. Pravst, *Br. J. Nutr.*, 2011, **106**, 1770.
24. EFSA, *EFSA J.*, 2008, **644**, 1.
25. European Commission, REFIT – Evaluation of the Regulation on nutrition and health claims, https://ec.europa.eu/food/safety/labelling_nutrition/claims/refit_en [accessed September 2018].
26. A. Kaur, P. Scarborough, S. Hieke, A. Kusar, I. Pravst, M. Raats and M. Rayner, *Eur. J. Clin. Nutr.*, 2016, **70**, 1462.
27. J. Aschemann-Witzel, N. Maroscheck and U. Hamm, *Food Qual. Prefer.*, 2013, **30**, 68.
28. H. Tuorila and A. V. Cardello, *Food Qual. Prefer.*, 2002, **13**, 561.
29. J. L. Orquin and J. Scholderer, *Food Policy*, 2015, **51**, 144.
30. L. Lahteenmaki, P. Lampila, K. Grunert, Y. Boztug, O. Ueland, A. Astrom and E. Martinsdottir, *Food Policy*, 2010, **35**, 230.
31. C. T. J. Lin, *Nutr. Today*, 2008, **43**, 267.
32. K. Brunsø, T. A. Fjord and K. G. Grunert, Consumers' food choice and quality perception, University of Aarhus, Aarhus School of Business, 2002.
33. E. Svederberg, *Forum: Qual. Soc. Res.*, 2002, **3**, 1.
34. Commission Regulation (EU) No 1169/2011, Official Journal of the European Union, 2001, L304/18.
35. EU, Commission implementing decision 2013/63/EU adopting guidelines for the implementation of specific conditions for health claims laid down in Article 10 of Regulation (EC) No 1924/2006, Official Journal of the European Union, 2013, L22/25.
36. (a) S. Storcksdieck genannt Bonsmann, L. Fernandez Celemin, A. Larranaga, S. Egger, J. M. Wills, C. Hodgkins and M. M. Raats, *Eur. J. Clin. Nutr.*, 2010, **64**, 1379; (b) FLABEL Project ID: 211905, Food Labelling to Advance Better Education for Life, https://cordis.europa.eu/project/rcn/88399_en.html [accessed September 2019].
37. W. L. Watson, B. Kelly, D. Hector, C. Hughes, L. King, J. Crawford, J. Sergeant and K. Chapman, *Appetite*, 2014, **72**, 90.
38. M. Raats, R. Malcolm, L. Lähteenmäki, I. Pravst, H. Gage, A. Cleary and M. Klopčič, *Nutr. Bull.*, 2015, **40**, 340.
39. REDICLAIM Project ID 63036, http://rediclaim.weebly.com/ [accessed September 2019].

CHAPTER 6

Measuring the Effects of Health Claims and Symbols: The CLYMBOL Methodological Toolbox

SOPHIE HIEKE* AND KLAUS G. GRUNERT

European Food Information Council (EUFIC), Brussels, Belgium
*Email: sophie.hieke@eufic.org

6.1 Introduction

In order to know how health claims and symbols affect consumer behaviour, the right methods need to be used to study such effects. As part of the EU-funded project CLYMBOL (Role of health-related claims and symbols in consumer behaviour, www.clymbol.eu, Project ID: 311963), a methodological toolbox was developed that contained scientifically validated, state-of-the-art methods for measuring how health claims and health symbols – in their context – are understood by consumers, and how they affect consumer food purchasing and consumption behaviours.[1]

For this purpose, we review existing literature on consumers and health claims and the methods applied evaluated based on set number of criteria. Those methods identified as the most promising were selected for further empirical testing regarding the types of effects studied in CLYMBOL, *i.e.* understanding, and effects on purchase and consumption. Such empirical testing was designed to allow comparison of the methods, both in terms of

Food Chemistry, Function and Analysis No. 22
Health Claims and Food Labelling
Edited by Siân Astley

Published by the Royal Society of Chemistry, www.rsc.org

theoretical anchoring and practical applicability. The objective has been to derive recommendations for the use of methods in studying the effects of health claims and symbols, for specific research questions, by specific stakeholder groups: research/academia, the food industry and policy-makers.

The toolbox was designed to offer guidance, *e.g.* for policy-makers and food business operators (FBOs) who need simple and robust, but scientifically sound, methods that they can apply routinely in testing claims and symbols for consumer understanding, and obtain summary insights concerning potential effects on purchasing and consumption. Furthermore, major policy shifts, such as the introduction of new health symbol schemes or revised legislation on the presentation of health claims, might call for more in-depth, policy-oriented methodologies that can capture their effects. Finally, methods are needed for scientific research in consumer science areas, usually with an emphasis on testing or identifying complex patterns of cause-and-effect behaviours.

6.2 Criteria to Assess Methods Measuring the Effects of Health Claims and Symbols

Three types of criteria were employed, dealing with:

1. Theoretical requirements resulting from theories on consumer understanding, purchase and consumption processes.
2. General criteria related to validity and reliability.
3. Feasibility (usability) in applying the methods to research questions.

6.2.1 Theory-driven Criteria

When studying consumer understanding of health claims, methods need to be able to deal with different situations in which consumers typically process health claims (*e.g.* shopping) and ways in which they infer meaning from health claims. Appropriate methods need to take into account how people form an understanding of health claims and symbols by relating them to their prior knowledge, and be able to measure inferences made manifest by the health claim. Lastly, the methods should have the potential to assess which inferences are justifiable and which are not, based on the scientific evidence provided to support the health claim.

With respect to the study of purchase behaviours, methods need to be able to track a purchase decision back to the attention given to individual elements on the packaging and be able to deal with antecedents of purchase decisions, such as attitudes or purchase intentions. Methods need to be able to deal with the continuum between habitual/heuristic and reasoned/deliberate choices. When studying consumption, methods need to be able to measure possible counter-effects that health claims and symbols might have both in terms of quantity and possible effects on overall composition of the diet.

6.2.2 General Criteria on Validity and Reliability

Methods should always be evaluated according to adequacy, *i.e.* capacity to examine the stated research question. Furthermore, face and content validity should be assessed, although the former is not always reported in empirical research. For quantitative studies, criterion and construct validity should be assessed, with the latter being the most important measure of validity. Reliability can be assessed using a range of tools, including internal consistency tests, test–retest reliability, and alternative forms of reliability testing. For qualitative research, although different in procedure, validity and reliability should still be assessed for each method used.

6.2.3 Feasibility in the Study of the Effects of Health Claims and Symbols

Considering both the range of research questions arising on consumer understanding, purchase and consumption effects as well as potential users of the toolbox, several feasibility criteria were developed by CLYMBOL to assess the suitability of a given method.

Methods should be suitable for the study of both verbal and symbolic claims as well as different types of health claims (*e.g.* function *vs.* reduction of disease risk claims) and symbols (*e.g.* wholegrain symbol *vs.* keyhole or choices logo). Methods should also be able to account for context (*e.g.* on pack or *via* additional means of communication), be suitable for use in different countries (*i.e.* different histories of health claims and symbols use) and different target groups as well as individual differences within the target groups.

More specifically, with respect to purchase, methods should be sensitive enough to pick up effects of interest and be able to measure not only reported but also actual behaviour. Concerning consumption effects, methods need to be able to deal with potential confounders and be suitable for cross-sectional as well as longitudinal studies and study of different product categories, and shed light on ways in which understanding, purchase and consumption are interrelated. Lastly, methods should be relatively simple and reasonably cheap to ensure applicability across a range of stakeholders (*e.g.* researchers *vs.* policy-makers).

For each of the effects (*i.e.* understanding, purchase and consumption), a literature review was undertaken to identify suitable methods, judged against the criteria listed. The selected methods were tested, through a series of experiments, to further determine suitability.

6.3 Methods for Testing Consumer Understanding of Health Claims and Symbols

Literature on consumer understanding generally distinguishes between objective and subjective understanding. Subjective understanding refers to how well consumers feel they understand a certain issue.[2] Objective understanding refers to whether the meaning consumers assign to a health claim or symbol and the inferences they draw from this are compatible with some objective benchmark of understanding.

Existing literature on inferences made from health claims has focussed mostly on a "magic bullet" effect (*i.e.* generalising to a broader range of health effects) and the halo effect (*i.e.* generalising positive perceptions to other product characteristics).[3] Other generalisations have been studied[4] and together, these types of inferences are a specific form of consumer misunderstanding.

Quantitative methods to measure consumer understanding included ratings (ratings, check-all-that-apply [CATA], ranking, and sorting), open-ended questions (open-ended questions, Consumer Understanding Test [CUT method], and thought-listing method), product comparison and match-up tasks, nutrient use accuracy tasks, and nutrient estimation tasks. Qualitative methods included interviews using specific techniques like laddering and focus group discussions. Based on the criteria defined previously, three methods were selected for measuring the inferences consumers make from health claims and symbols, specifically the CUT method, and soft and hard laddering interviews.

The criteria used to select these methods were:

- Sensitivity to differentiate between health claims
- Capacity to compare inferences to the health claim dossier
- Consider how people relate the health claim to prior knowledge
- Richness of data gathered and extent of unusable data
- Cost
- Ease of use
- Usability of results for the main stakeholder groups

Three health claims, related to vitamin D, salt reduction and beta-glucan, were selected based on relevance to public health and degree of familiarity with the nutrients and functions amongst the population.† Three carrier products (fish, yoghurt and bread) were also selected such that the health claims might fit with at least two out of three of these products. The health claims were presented in different forms, namely as the authorised claim or as a reformulated version of that claim, with or without an additional general health claim, and with or without additional information explaining the terms in the authorised claim (for details see ref. 5).

6.3.1 Consumer Understanding Test (CUT Method)

CUT was developed by Danone, specifically to measure understanding of health claims.[6] Based on two questions, "If you had to tell a friend what

†The design of this study, and especially the selection of health claims, carrier products, and the different versions of the health claims, were carried out in cooperation with the Danish Meal Partnership, a stakeholder roundtable organization to promote healthy eating.

[this fruit yoghurt/this mackerel in tomato sauce/this wholegrain bread] does, what would you say?'' and "And, if you had to tell your friend how it works?'', open-ended answers for all three carrier products were analysed and coded as three types of inferences: "safe'' (*i.e.* correct interpretation), "risky'' (*i.e.* incorrect interpretation) and "other'' (*i.e.* vague, neither correct nor incorrect interpretation). The scientific dossier for each was used to decide whether a response was safe, risky or other.

Most inferences were classified as vague. Safe inferences were much less prevalent, but still more frequent than risky inferences (*i.e.* incorrect interpretation). A separate analysis by claim type revealed differences in how well each of the different claims were understood. CUT was well suited to identifying differences in understanding between the health claims and the wordings and context of each claim. This method provided a clear answer to the question: "to which extent do consumers understand a particular health claim?''

6.3.2 Soft Laddering Interviews (Face-to-Face)

Laddering is an interview technique designed to elicit series of inferences and is therefore better suited to establishing insights not only into whether consumers' inferences are correct compared with the scientific dossier, but also how these inferences come about. In face-to-face interviews, a ranking of three products from one of the three selected categories was carried out. Two of the three products had one of the versions of a health claim on them.

Respondents were told, "Imagine you are out shopping, and you see the following three products, of which some – as it is shown in the picture – have some extra information on the package'', before they were asked to rank the stimuli products according to the likelihood that they would choose them. In a second step, they had to provide reasons for their ranking. In a third step, respondents were asked to give reasons for their rankings that was followed up with a series of "why is it important to you that the product. . ..'' questions followed "and why is it important to you that. . .'', which is the basis of laddering.[7] With this method, it is possible to compare inferences consumers make from the different health claims and the context in which they appear. For example, adding a general health claim to the specific health claim diverts consumer attention, and leads to a range of more general associations that are not in accordance with the scientific dossier.

6.3.3 Hard Laddering (Online)

As soft laddering is face-to-face and, therefore, expensive, a hard-laddering method was also investigated. An online version of laddering was implemented, where respondents, after the ranking task, had to give (write) reasons for the ranking in three hierarchically-ordered boxes. The results showed that, compared with soft laddering, fewer inferences were measured, but the method still offered a suitable approach for identifying differences in how consumers interpreted different wording and context of a health claim. These methods allow researchers to draw a more comprehensive picture of the thinking process that takes place when consumers interpret a health claim.

6.3.4 Conclusions

All three methods were evaluated with regard to their usability for different stakeholders. The methods were sufficiently sensitive to differentiate between different claims/wordings and took account of how people relate claims to their prior knowledge. CUT was best for comparing inferences made to evidence in the scientific dossier. Soft laddering produced the richest data and the least noise (*i.e.* unusable data) but is the most expensive and difficult to use due to study design and coding. Both laddering methods were well suited for tracking the inference process in the mind of consumers confronted with health claims.

6.4 Methods for Testing Effects of Health Claims and Symbols on Consumer Purchases

Generally, purchasing behaviour is defined as a process that starts with problem recognition and extends *via* information search, product evaluation, decision-making and actual purchase to the post-purchase phase.[8] In measuring effects of health claims on purchase, it is important to analyse not only effects on purchase, but also in the stages of purchasing preceding the act. In particular, it is important to know whether and how the health claim attracts attention, as this is a prerequisite for the health claim being part of the information search process, whether health claims lead to arousal (facilitating approach behaviour and information processing), and if health claims lead to a positive attitude towards the product.

A number of quantitative methods to measure purchase behaviour were reviewed, including use of sales and scanner data, panel data, self-reported behaviour and buying intentions, as measured in a questionnaire, virtual shelves and experimental product choice. Electrodermal reaction was identified as a way to measuring arousal, and eye-tracking as a way to measure attention. Explicit and implicit to measure attitudes were also reviewed.

Qualitative methods included focus group discussions and observations. Four methods were selected for testing, namely eye-tracking (attention), electrodermal activity (arousal), implicit measures (attitude) and experimental choice. Actual sales data were also collected, as a benchmark against which the other measures could be compared. The criteria used to select these three methods were (1) sensitivity to differentiate between different types of health claims and product categories, (2) ability to account for context factors, (3) sensitivity for individual differences, (4) applicability across countries, and (5) suitability for cross-sectional as well as longitudinal studies. Cost and ease of use differed considerably, which impacts their usability for stakeholders.

Three olive oils labelled with different claims were used for this study, including a taste claim (*fruity and savoury*), a nutrition claim (*high in vitamin E*) and a nutrition–health claim combination (*High in Vitamin E: Vitamin E contributes to the protection of cells from oxidative stress*).

6.4.1 Sales Data

Sales data, as a measure of actual purchase behaviours, were collected in a two-week field experiment at a GLOBUS supermarket in Germany (beneficiary in the CLYMBOL project). Packaging for the in-house brand of olive oil was amended to carry the three different claims. The price was unaltered, and all promotional activities for that product were controlled during the timeframe of the experiment. Sales data revealed that the highest proportion of GLOBUS olive oils sold carried a nutrition claim (*i.e. high in vitamin E*), followed by the nutrition–health claim combination and, then, the general taste claim.

6.4.2 Eye-tracking Study

Using the sales data as a point of reference, an eye-tracking study was carried out where participants were given a shopping list that included olive oil and, after calibration of the eye-tracking devices, were asked to complete their shopping task in an experimental supermarket. Nothing was purchased, but participants completed the choice task, meaning they had to select one product from each category on their shopping list. Attention to a claim on the product was measured by fixation (at least one fixation on the claim), gaze duration (sum of fixations on the claim) and gaze frequency (how often the claim was looked at) (for an introduction to the use of eye-tracking in analysing attention to information on food packages, see ref. 9).

Results showed that participants were more likely to pay attention to the nutrition claim compared with the nutrition–health claim combination. Overall, the odds of fixing a claim were 5.5 times higher (significant at $P<0.05$) for the nutrition claim than the nutrition–health claim combination. Across all conditions, those that had at least one fixation on the claim were significantly more likely to choose the olive oil carrying a nutrition claim. These results indicate not only that eye-tracking is useful in analysing whether a claim attracts attention, but also that attention, as measured by eye-tracking, is predictive of choice behaviour.

6.4.3 Electrodermal Reactivity Study

In order to measure arousal (approach behaviour), electrodermal reactivity was measured in a separate purchase simulation study where participants were asked to select three products, including olive oil, from a virtual shelf (for more information on the role of arousal in shopping and its measurement by electrodermal reactivity see ref. 10). Instructions were the same as the eye-tracking study, with the only difference being that they were seated in front of a computer, with electrodes attached to their left hand, rather than in an experimental supermarket when making their choices. Measures of arousal included amplitudes (strength of each arousal action), total amplitude (sum of all amplitudes) and frequency (number of all single reactions). Results showed that a nutrition or health claim increased arousal (sum and frequency of amplitudes) significantly, with the nutrition claim leading to greater arousal (sum amplitude) than a health claim. Whilst arousal was highest for the nutrition claim condition, claim type did not impact whether participants chose the olive oil compared with four options without a claim, *i.e.* arousal is not predictive of choice behaviour.

6.4.4 Implicit Attitude Study

Implicit attitudes were measured using a reaction-time task. Response latencies (reaction times), measured for categorising a noun as positive or negative, immediately after participants were shown images of the olive oil bottles, were used as a proxy for attitude (for an introduction to the use of implicit attitude measures in consumer research, see ref. 11). Results showed that the most positive implicit attitude was associated with the taste claim. Participants needed 15 ms less to categorise a positive target, compared with negative targets, after exposure to the olive oil bottle carrying a taste claim. This effect was marginally significantly different for the effect of both the nutrition claim and the nutrition–health claim.

6.4.5 Conclusion

Ultimately, whilst actual sales data measured the effects of health claims and symbols on consumer purchase behaviour, such data are only available once a product is on the market. However, proximal measures can also provide insights into why a purchase occurs or not. Paying attention to a health claim or symbol on the packaging is a prerequisite for claims/symbols to have an effect on purchase. Eye-tracking was a useful method for analysing which claims attracted attention and how much. Results from the study carried out by CLYMBOL indicate that using eye-tracking allows differences in the attention-getting properties of different claim types to be detected and, thus, help explain purchase behaviour. This method is, however, very time- and resource-intensive, and requires expertise in carrying out this type of study and analysing the data. Arousal, as measured by electrodermal reactivity, was also able to detect differences between claim types by revealing unconscious approach behaviour, although the relationship between arousal and choice was weak. Implicit attitude measures are only useful in measuring attitude once consumers have paid attention to a claim. CYMBOL also showed that simulated choices, regardless of the environment (experimental supermarket with real products *vs.* a virtual shelf) were well aligned with actual choices and are, therefore, a useful approach for measure of purchase behaviour for products not yet on the market.

6.5 Methods for Testing Consumption Effects of Health Claims and Symbols

At an individual level, consumption is a complex process influenced by a number of physiological and psychological factors. For any consumption study, in order to choose the right method, it is important to decide in advance which indicators are to be measured. Pre-consumption measures, for example, include intention or expectation to consume. Actual consumption behaviour can be measured through observed or self-reported methods. Post-consumption measures include nutrient intake and food energy and/or energy balance. Many other factors, such as sensory properties of food or physiological state, can influence consumption as well but were not included in this overview, as they may be insufficient as proxies when studying the effects of claims and symbols.

A number of methods were reviewed, covering all stages of consumption. For pre-consumption, methods included food balance sheets, as a measure of food availability, and household surveys. Actual consumption measures included observation, dietary records, food checklists, 24-hour recall, food frequency questionnaires and diet history interviews. Post-consumption measures included biochemical markers and anthropometry. As each method can only capture part of the overall consumption process and, therefore, also the (potential) effect that a claim-carrying product can have, a study was designed involving a number of different measures. Methods were chosen that would capture both quantity and quality of food intake, employing both observation and self-reporting, and covering both prospective and retrospective food intake. More specifically, repeated real-world food diaries were combined with one-off laboratory recordings of food weight, before and after consumption. The validity of the food diaries was measured using urine samples as well as changes in weight or energy balance.

6.5.1 The Breakfast Study

The study protocol followed a single-blinded, parallel, randomised, experimental design. Participants completed three visits as well as an at-home completion of food and physical activity diaries, and urine samples. At the first visit, participants were asked to fill in a welcome questionnaire, were measured (height, weight, and hip-to-waist ratio), and asked to wear an activity watch, recording sleep and activity over the duration of the study. Then, they completed a four-day food and physical activity diary, and (males only) collected two 24 h urine samples. At the second visit, participants were randomised into a health claims or control group. This visit was introduced as a fasted breakfast taste test. Containers of cereal carrying either a health claim (*Oat beta-glucan has been shown to lower blood cholesterol*) or a control label were displayed throughout the session. Participants were asked to complete a hunger questionnaire and then taste and rate the cereal, followed by *ad libitum* cereal consumption (as wished for breakfast). Participants were provided with a bag of cereal (carrying either a health claim or a control label) and asked to continue to eat the cereal at home (to increase exposure to the product, either with or without the health claim) and complete a second four-day food and physical activity diary, and (males only) collect two further 24 h urine samples. The third visit included completion of an exit questionnaire, repeated anthropometric measurements, a compliance interview, and an explanation of the study purpose.

Results from the baseline four-day food and drink diaries revealed no significant differences between the health claim and the control group, regarding macro- or micronutrients consumed. Health claim exposure did not result in any significant change in dietary consumption of the cereal in the laboratory setting. As for the baseline, results from the one-day dietary intake recorded on the day the participants attended the laboratory tasting session revealed no significant differences in dietary intake of macro- or micronutrients between health claim exposure and control groups. The only exception was protein intake, where 35% more of the recommended dietary reference value was consumed by the control group compared with the health claim group. Results from the second four-day food and drink diaries (post-intervention) also revealed no significant differences between the health claim and the control groups. Within groups, a comparison of micro- and macronutrient intakes over time found minimal differences in intake between baseline and post-intervention. Between groups, vitamin B12 intake was the only macro- or micronutrient where intake between groups differed significantly over the course of the study (intake increased in the control group and decreased in the health claim group).

6.5.2 Conclusion

Despite elaborate design, the Breakfast Study did not show any statistically significant differences in consumption of a product with or without a health claim. The experiment did, however, show that it is possible to combine laboratory and field measurements to assess consumption effects. The literature reports widely that dietary consumption methods can lead to undetectable effects because of underlying limitations. This is reflected in health claims research where consumption *per se* is not the outcome measured due to difficulties in such measurements. One question to ask when designing such a study could be whether purchase methods are adequate proxies for food choice. Thus, methods are needed that allow for the study of dietary patterns and not just nutrient content including validated, health-claim-sensitive dietary or food frequency questionnaires. In the meantime, and perhaps even with these methods, it is advisable to use more than one method to measure consumption in order to corroborate any findings.

6.6 Recommendations for the Use of Methods by Various Stakeholders

Building on the review of methods and the studies presented above, the following recommendations have been derived for the use of methods by various stakeholders.

6.6.1 Recommendations for Regulators

An important question for regulators is whether a claim is understood correctly by the average consumer. A combination of CUT and laddering methods is recommended. CUT will show whether the claim is understood correctly whilst laddering can show why some consumers might be misled by the claim, by allowing inference processes to be followed. This combination is also useful to evaluate whether a health claim or symbol will lead to healthier choices. CUT will demonstrate understanding whilst laddering will analyse how changing the wording influences consumer understanding. To observe the influence of health claims on consumer choice, a combination of eye-tracking and choice experiments is advised. Eye-tracking will show how much attention a claim receives, and the actual choice can be evaluated using choice experiments, as purchase data might not be available.

6.6.2 Recommendations for the (Food) Industry

Food manufacturers should use CUT to document whether a claim or symbol is understood as required by Regulation (EC) 1924/2006. This method not only analyses whether a claim is understood correctly, but also has the advantage of being readily applicable to larger samples representative of the general population or a specific target group. Furthermore, CUT can pick up the influence of other factors, such as on-pack marketing slogans. However, presumably, for the (food) industry, the most decisive factor is the influence of a claim or symbol on purchase. In order to trigger consumer choice, a claim must first catch the attention of customers. Eye-tracking is a reliable and valid method to evaluate the capacity of product information to attract attention. This method allows testing of different versions and contexts of a claim, in a laboratory and/or supermarket. To test whether a health claim influences brand or company image, laddering analyses inferences consumers make from a product and information it carries. The perceived image of a product might not lead to an immediate purchase but may influence future decisions.

6.6.3 Recommendations for Consumer Scientists

Similar to the recommendations above, researchers should use both CUT and laddering methods to evaluate consumer understanding. A set of different methods is required to evaluate purchase behaviours. Eye-tracking can be used to measure attention and spontaneous consumer reactions to the product can be evaluated using electrodermal reactions to measure arousal. However, results suggest that the spontaneous effects resulting from health claims and symbols are limited. Choice experiments are recommended to test consumer purchasing, as sales/scanner data might not always be available. The limitations of consumption methods are widely known. Therefore, researchers need to consider various factors in selecting the most accurate and reliable method that is also feasible and appropriate for the research question in order to help inform study design.

Recommendations from CLYMBOL for the future design of studies into health claims include considering whether consumption is the outcome measure of choice (*i.e.* is it possible or necessary to have a consumption measure that can identify health-claim-relevant consumption or are purchase methods adequate?). If consumption must be considered, then it needs to be defined (*i.e.* does consumption refer to the amount consumed or the frequency of consumption or both?). Alternatively, will the outcome measured be food choice, intention to consume, or intake measured at the dietary, food or nutrient level? Whatever the answers, single-item questionnaire survey measures are not likely to be sufficient. Instead, researchers should be aware of their limitations and consider the potential of incorporating more than one measure of consumption in their study designs.

References

1. S. Hieke, N. Kuljanic, J. M. Wills, I. Pravst, A. Kaur, M. M. Raats, H. C. M. van Trijp, W. Verbeke and K. G. Grunert, The role of health-related claims and health-related symbols in consumer behaviour: Design and conceptual framework of the CLYMBOL project and initial results, *Nutr. Bull.*, 2015, **40**, 66–72.
2. H. C. Van Trijp and I. A. Van der Lans, Consumer perceptions of nutrition and health claims, *Appetite*, 2007, **48**, 305–324.
3. B. Roe, A. S. Levy and B. M. Derby, The impact of health claims on consumer search and product evaluation outcomes: Results from FDA experimental data, *J. Public Policy & Marketing*, 1999, **18**, 89–105.
4. J. C. Andrews, R. G. Netemeyer and S. Burton, Consumer generalization of nutrient content claims in advertising, *J. Marketing*, 1998, **62**(4), 62–75.
5. V. Stancu, K. G. Grunert and L. Lähteenmäki, Consumer inferences from different versions of a beta-glucans health claim, *Food Qual. Preference*, 2017, **60**, 81–95.
6. M. Rogeaux, Consumer understanding and reaction to health claims. Insights and methodology, in *Consumer-driven Innovation in Food and Personal Care Products*, ed. S. R. Jaeger and H. MacFie, Woodhead, Cambridge, 2010, pp. 277–302.
7. K. G. Grunert and S. C. Grunert, Measuring subjective meaning structures by the laddering method: Theoretical considerations and methodological problems, *Int. J. Res. Mark.*, 1995, **12**, 209–225.

8. K. G. Grunert, Consumer behaviour, in *The Oxford Textbook of Marketing*, ed. K. Blois, Oxford University Press, Oxford, 2000, pp. 109–130.
9. D. J. Graham, J. L. Orquin and V. H. Visschers, Eye tracking and nutrition label use: A review of the literature and recommendations for label enhancement, *Food Policy*, 2012, **37**, 378–382.
10. A. Groeppel-Klein, Arousal and consumer in-store behaviour, *Brain Res. Bull.*, 2005, **67**, 428–437.
11. D. Maison, A. G. Greenwald and R. H. Bruin, Predictive validity of the Implicit Association Test in studies of brands, consumer attitudes, and behavior, *J. Consum. Psychol.*, 2004, **14**, 405–415.

CHAPTER 7

Food Allergen Labelling Regulation

LUCA BUCCHINI,*[a] MATTHEW DALY[b] AND
E. N. CLARE MILLS[b]

[a] Hylobates Consulting srl, via Tarsia 61, 00135 Rome, Italy; [b] Division of Infection, Immunity and Respiratory Medicine, School of Biological Sciences, Manchester Academic Health Sciences Centre, Manchester Institute of Biotechnology, Manchester M1 7DN, UK
*Email: lucabucchini@hylobates.it

7.1 Introduction

Food allergies have risen in prominence over the last 20–30 years and currently, as there is no accepted cure, individuals usually have to practice life-long avoidance of their problem food(s). There are many different types of food allergy and intolerance, but those involving the immune system – either IgE-mediated food allergies or the T cell-mediated adverse reaction to gluten (*i.e.* coeliac disease) are amongst the most important. For those with IgE-mediated allergies, who are at risk of severe reactions, avoiding problem foods – even in trace amounts – is vital. Although rare, individuals with such allergies can be hospitalised as a consequence of accidental exposure and there is risk of a fatal reaction (anaphylaxis), even if it is rare. For individuals with coeliac disease, avoiding gluten effectively is equally important for their long-term health and well-being, although the effects of accidental gluten consumption are not as acute. Allergic consumers need to identify, without doubt, products that are safe for them to eat as well as those that should be avoided. This requires clear indication of the presence of certain foods and

Food Chemistry, Function and Analysis No. 22
Health Claims and Food Labelling
Edited by Siân Astley

Published by the Royal Society of Chemistry, www.rsc.org

derived ingredients on product labels. Although other types of intolerance (*e.g.* enzyme deficiency that leads to lactose intolerance) also require labelling strategies to help consumers manage their diet, this review focuses on the food labelling of allergens that seeks to protect those with immune-mediated allergies. Throughout this chapter the term allergen is used to indicate foods of concern in a food industry or food safety context, although this has different meanings in scientific and clinical domains, where it refers to allergenic protein molecules – whole or partial – in a food.

7.2 Food Allergen Regulation and Labelling

7.2.1 Mandatory Allergen Labelling

Food ingredients lists have been mandatory for prepackaged foods for many years, the CODEX General Standard for labelling of prepackaged foods having been adopted in 1969. However, under the original standard neither minor ingredients added to a recipe or compound ingredients (made up of two or more other ingredients) had to be declared. As a consequence, allergenic ingredients, such as cow's milk used to carry flavours, were not listed as ingredients, despite posing a risk to food-allergic consumers.[1] Early in the 1990s, there was a general consensus that individuals with food allergies should be able to consume safely foods intended for the general population and, consequently, in 1991, Norway raised the issue formally in the context of the CODEX Alimentarius Commission.[2] However, alternative standpoints existed and, as late as 1997, some countries resisted reforms championed by the Nordic countries, advocating that consumers should obtain information from resources besides labelling, such as the Dutch Food Intolerance Databank.[3] Another approach that could have been adopted for consumers with IgE-mediated food allergies was that developed for individuals with other immune-mediated adverse reactions to food, *i.e.* coeliac disease. In Europe at that time, people with coeliac disease were provided with foods classified as being "for particular nutritional uses (PARNUTS)", which were regulated by Council Directive 77/94/EEC. However, such products would have been more expensive and difficult for allergic consumers to find and did not present a solution to the public health problem posed by either IgE-mediated food allergies or coeliac disease.

It took almost 10 years for CODEX to adopt a list of foods of allergenic concern that should be labelled irrespective of their level of inclusion and, thereby, disallowing previous exemptions for compound ingredients (ingredients made up of two or more other ingredients). The first step in this process was to develop a list of foods of allergenic concern. Although hundreds of foods have been described as sensitising individuals or triggering allergic reactions, only a few are considered to pose a public health risk and are included in the legislation. Thus, in 1999, the CODEX Alimentarius Commission adopted the "General standard for the labelling of prepacked foods", which included a list of priority allergenic foods

Table 7.1 CODEX Alimentarius Commission (1999) list of priority "foods and ingredients known to cause hypersensitivity".

Cereals containing gluten; *i.e.* wheat, rye, barley, oats, spelt or their hybridised strains and products of these
Crustacea and products of these
Eggs and egg products
Fish and fish products
Peanuts, soybeans and products of these
Milk and milk products (lactose included)
Tree nuts and nut products
Sulphite in concentrations of 10 mg kg^{-1} or more

Table 7.2 Timeline for the development of allergen labelling regulations across the world.

Year	Country/bloc/international organisation	Event
1991	CODEX Alimentarius Commission	Norway proposes a paper on Labelling of Potential Allergens in Foods
1995	CODEX Alimentarius Commission	FAO Technical Consultation on Food Allergies takes place
1999	CODEX Alimentarius Commission	CODEX adopts new rules for foods known to cause hypersensitivity
1999	Japan	Food Sanitation Law
2002	Australia and New Zealand	Australia New Zealand Food Standards Code Standard 1.2.3 introduced[8] (mandatory labelling)
2003	EU	Directive 2003/89/EC (food allergen directive) enacted
2004	USA	FALCPA enacted
2007	EU	Directive 2007/68/EC (last modification of allergen list) enacted
2011	EU	Regulation 1169/2011/EU (incl. non-prepacked food and catering) enacted[9]
2012	Canada	Food and Drug Regulations Act in force
2012	China	National Food Safety Standard on General Rules for the Labeling of Prepacked Foods (GB7718-2011) enters into force[10]
2013	Russian Federation	Customs Union Technical Regulations on Food Products Labeling enters into force[11]
2013	Japan	New guidance on food allergens
2016	Association of Southeast Asian Nations (ASEAN)	General standard for the labelling of prepacked foods endorsed[12]

(Table 7.1).[4] The list comprised eight allergenic foods[5] and has served as a starting point for priority allergenic foods in regulations in many countries and regions of the world (Table 7.2). Although not an allergen, sulphites were also included in this list because of their capacity to trigger adverse reactions similar to those involved in IgE-mediated food allergies, although the underlying mechanisms are different.

In 2009, Japan was the first country to implement allergen labelling, although with a rather different approach to the mandatory allergen lists.[6] Adoption of CODEX recommendations in the European Union (EU) followed when a major revision of EU labelling law was undertaken in 2000, enacted through Directive 2003/89/EC.[7] The original CODEX list was elaborated with other foods considered to be of allergenic concern within EU Member States (MS), such as mustard and celery, and provided for exemptions and updating of the priority allergen lists in the future. The USA followed suit in 2004 with the Food Allergen Labelling and Consumer Protection Act (FALCPA). Thus, the 1990s were a decade that raised awareness of food allergies and the need for regulation, but adoption across the world has taken much longer (Table 7.2).

The current decade (2010–20) has seen the introduction of allergen regulations globally, including China, Brazil, Argentina, the Russian Federation and ASEAN (Association of Southeast Asian Nations, comprising Brunei, Darussalam, Cambodia, Indonesia, Laos, Malaysia, Myanmar, Philippines, Singapore, Thailand and Vietnam). Many countries have simply endorsed the CODEX list of foods (for example, ASEAN and China) whilst others – specifically Argentina, Australia and New Zealand, Brazil, Canada, the EU, Japan, Russia, Switzerland and the USA – have added to this list. This decade has also seen some regions and countries improve allergen labelling legislation further. For example, in the EU, the loophole for compound ingredients was closed in 2011, and presentation of allergen information on food labels was regularised through Regulation (EU) No 1169/2011. In 2011, the EU also extended some labelling provisions to non-prepacked food, such as meals served in hotels, restaurants and other catering outlets and, although fraught with challenges, other regions might follow in the near future. Improvements have also been made in Canada (2012) and Australia and New Zealand (2017).

7.2.2 Regulated Priority Food Allergens around the World

Although foods are identified in the legislation using different terminology (*e.g.* major allergens in the USA or substances or products causing allergies or intolerances in the EU), it is generally agreed these encompass ingredients, including processing aids produced or derived from such foods, unless exempted; processing aids are substances used intentionally in the processing of raw materials, foods or their ingredients, to fulfil a specific technological purpose during treatment or processing that could result in their unintentional but technically unavoidable presence in the final product; typical processing aids are food enzymes. However, this does not apply to unintended allergen presence, as a consequence of allergen cross-contact resulting from the use of shared equipment and processing lines. According to the US Food and Drug Administration,[13] FALCPA sets out specific requirements for allergen labelling that apply to any "food ingredient that contains protein derived from a food", which includes flavorings, colorings, or incidental additives (*e.g.* processing aids). Similarly, in the EU, Regulation (EU) No 1169/2011[9] does not exempt from food allergen labelling requirements processing aids, carriers or any other substance that might usually be omitted from the list of ingredients (article 20), unless a derogation has been granted specifically.

The structure of the regulations varies with some stipulating derived products in the allergen lists (as done by the EU) whilst others define these as outside the allergen priority list. The original CODEX priority list of allergenic foods still forms the basis for almost all legislation around the world, but there are many variations in the lists, as countries try to take account of the needs of their particular population or interpret foods that should/should not be distinguished specifically. For example, like many countries in South America, China does not distinguish between crustacean and molluscan shellfish in the National Food Safety Standard on General Rules for the Labeling of Prepacked Foods lists the eight foods in line with CODEX (Table 7.3).[10]

In July 2015, Brazil introduced mandatory labelling of food allergens with Resolução – rdc n° 26, of 2 July 2015, which lists 18 foods that cause food allergies. Milk from all mammalian species is included as well as nine (tree) nuts; natural latex is also included. Resolución Conjunta 57/2010 y 548/2010 modified the Argentinian Food Code to incorporate 10 food allergens, including the food colour tartrazine (E102). In addition to the CODEX list, the Australian and New Zealand Food Standards (FSANZ) Code Standard priority allergen list includes sesame and lupin, but not molluscan shellfish or mustard.[14] It is interesting to note that under Standard 1.2.3-3, royal jelly requires a warning stating it "...has been reported to cause severe allergic reactions and, in rare cases, fatalities, especially in asthma and allergy sufferers", but is not otherwise treated like a food allergen. In the EU, Regulation (EU) No 1169/2011[9] includes a list of 14 "substances or products causing allergies or intolerances" in Annex II, with celery unique in Europe as a food of concern. The last modification to the list occurred in December 2006, with the addition of molluscs and lupin. The same list is used in Switzerland, under the relevant Ordnance of 23 November 2005. Since 2013, the Russian Federation[11] included 14 allergens and, globally, is the most similar to the EU list.

In the USA, FALCPA lists the CODEX eight major allergens, but does not include molluscs,[13] cereals containing gluten, other than wheat (such as oats), or allergens identified in other lists. It also requires the types of tree nut (*e.g.* almonds, pecans, walnuts), fish (*e.g.* bass, cod) and crustacean shellfish (*e.g.* crab, lobster, shrimp) to be declared. The emphasis in FALCPA is placed on the protein fractions of major allergenic food sources and, providing there is evidence that proteins are not present (*e.g.* in highly refined oils), the allergenic ingredient does not have to be declared. Food allergen legislation in Canada also specifies that food allergens relate exclusively to the protein fraction of a food, including any modified proteins.[15] Both molluscan and crustacean shellfish are listed, together with sesame

Table 7.3 Allergenic foods and derived ingredients which require labelling to protect allergic consumers in different regulatory jurisdictions. All regulatory authorities require that sulphites are labelled if used at a concentration greater than 10 mg kg^{-1}, apart from Brazil. Brazil also requires natural rubber latex to be declared. Allergen labelling in Russia is voluntary. Japan has a second list of food allergens for voluntary labelling, which includes molluscan shellfish species (abalone, squid), fish (salmon roe, salmon, mackerel), fruit (bananas, apples, oranges, peaches, kiwi fruit), tree nuts (cashew nut, walnuts), and seeds (soybeans, sesame), and meat (beef pork, chicken and gelatin), together with mushroom, yams and matsuke. Hickory nuts are listed by US FALCPA regulations and, generally, come from *Carya ovata*, also known as shagbark hickory, although the species is not listed in the Regulation.

CODEX	Hong Kong	China	USA	Australia and New Zealand	Canada	EU	Russia	Japan	Brazil
Cereals containing gluten, *i.e.* wheat, rye, barley, oats, spelt or their hybridised strains and products of these	Cereals containing gluten, *i.e.* wheat, rye, barley, oats, spelt or their hybridised strains and products of these	Wheat	Wheat	Cereals containing *gluten, namely, wheat, rye, barley, oats and spelt and their hybridised strains	Wheat or triticale and gluten	Cereals containing gluten, namely: wheat (such as spelt and Khorasan wheat), rye, barley, oats or their hybridised strains, and products thereof	Cereals containing gluten, and their products	Wheat	Wheat, rye, barley and oats
Crustacea and products of these	Crustacea and crustacean products	Shellfish	Crustacean shellfish (*e.g.* crab, lobster or shrimp)	Crustacea	Crustaceans	Crustaceans and products thereof	Crustaceans and their products	Shrimp and crab	Crustaceans
Eggs and egg products	Eggs and egg products	Egg	Egg	Egg	Eggs	Eggs and products thereof	Eggs and their products	Egg	Eggs
Fish and fish products	Fish and fish products	Fish	Fish (*e.g.* bass, flounder or cod)	Fish	Fish	Fish and products thereof	Fish and its products	Fish	Fish

Milk and milk products (lactose included)	Milk and milk products (including lactose)	Milk	Milk	Milk	Milk	Milks and products thereof (including lactose)	Milk and its products (including lactose)	Milk	Milk
Peanuts and products of these	Peanuts and peanut products	Peanuts	Peanuts	Peanuts	Peanuts	Peanuts and products thereof	Peanuts and its products	Peanuts	Peanuts
Soybeans and products of these	Soybeans and soybean products	Soybean	Soybeans	Soybeans	Soybeans	Soybeans and products thereof	Soybean and its products	—	Soybeans
Tree nuts and nut products	Tree nuts and nut products	Tree nuts	Tree nuts (*e.g.* almonds, pecans or walnuts)	Tree nuts	Almonds, Brazil nuts, cashews, hazelnuts, macadamia nuts, pecans, pine nuts, pistachios or walnuts	Nuts, namely: almonds (*Amygdalus communis* L.), hazelnuts (*Corylus avellana*), walnuts (*Juglans regia*), cashews (*Anacardium occidentale*), pecan nuts (*Carya illinoinensis* (Wangenh.) K. Koch), Brazil nuts (*Bertholletia excelsa*), pistachio nuts (*Pistacia vera*), macadamia or	Nuts and their products	—	Almonds, Brazil nuts, cashews, hazelnuts, macadamia nuts, pecans, pine nuts, pistachios, walnuts and chestnuts

Table 7.3 (*Continued*)

CODEX	Hong Kong	China	USA	Australia and New Zealand	Canada	EU	Russia	Japan	Brazil
						Queensland nuts (*Macadamia ternifolia*) and products thereof			
Additional foods not included in the CODEX list									
				Sesame seeds	Sesame seeds	Sesame seeds and products thereof	Sesame and its products		
				Lupin		Lupin and products thereof	Lupin and its products		
				Not required	Shellfish	Molluscs and products thereof	Molluscs and their products		
				Not required	Mustard seeds	Mustard and products thereof	Mustard and its products		
				Not required		Celery and products thereof	Celery and its products		
								Buckwheat	

and mustard, but the Canadian list does not include lupin. The other unusual aspect of Canadian allergen legislation is that wheat and triticale are included whilst gluten is mentioned separately and defined as the protein fraction of barley, wheat, triticale, rye and oats. The Japanese regulatory system is also unusual in that it has only seven mandatory food allergens, including buckwheat, based on local prevalence. However, labelling is recommended for a further 18 allergens including latex, a non-food allergen.[6]

It is worthwhile noting that tree nuts and cereals, because they are diverse groups, have resulted in widely different interpretations; whilst many lists lack detail, others are very specific (Table 7.3). In the case of nuts, for example, CODEX was non-specific, referring simply to *tree nuts*, as opposed to ground nuts (*i.e.* peanuts). The EU list indicates botanical species, but is more restrictive than the FALCPA, which – in addition to the three examples (almonds, pecans and walnuts) – includes a longer list of tree nuts[16] (Table 7.3). FSANZ legislation exempts explicitly "coconut from the fruit of the palm *Cocos nucifera*" whilst, in Canada, tree nuts include pine nuts. Thus, food business operators (FBOs) and consumers need to be aware of the longer US list, for example, when trading raw materials or purchasing products containing tree nuts because some ingredients will not have been included on labels in the EU (*e.g.* pine nuts). The EC has stated that the EU list is exhaustive, meaning other foods will not be added in the foreseeable future.[17]

Such diversity in priority food allergen labelling lists means that, in ensuring compliance, the first requirement for FBOs (manufacturers, suppliers) is to understand which allergen labelling applies to which foods. This is crucial because allergens might require specific management in the supply chain and processing environment as well as appropriate labelling. For example, coconut is designated as an allergen in the USA but not in other jurisdictions. Thus, a company based in the EU can treat coconut powder as an ordinary ingredient (*i.e.* not an allergenic food), provided products are not marketed in the USA. However, at some later stage, the FBO might decide to export coconut-containing products to the USA. In that case, those responsible for quality and regulatory affairs must ensure that evidence from suppliers, which forms part of labelling decision processes, includes information about coconut-derived ingredients or processing aids that may be omitted in EU technical specifications and cross-contact risks. This process might result in the label having to be updated to include coconut, with the associated costs.

There are also subtle differences in definitions around the world relating to gluten (Table 7.3). In line with CODEX, the EU Regulation includes "cereals containing gluten" with a specific list of cereals "and their hybridised strains", with one notable exception, Khorasan wheat. The original EU list stated Kamut®, but this was corrected to *Khorasan wheat*, as Kamut® is a registered trademark. China and Russia have similar legislation to the EU, but the situation is different in the USA. The US Food and Drink Administration (FDA)[16] clarified the term *wheat* in Section 201(qq) and included all species of the genus *Triticum*, such as common wheat (*Triticum aestivum* L.), durum wheat (*Triticum durum* Desf.), club wheat (*Triticum compactum* Host.), spelt (*Triticum spelta* L.), semolina (*Triticum durum* Desf.), Einkorn (*Triticum monococcum* L. subsp. *monococcum*), emmer (*Triticum turgidum* L. subsp. *dicoccon* (Schrank) Thell.), Kamut (*Triticum polonicum* L.) and triticale (x *Triticosecale* ssp. Wittm.). Barley, rye and oats were not included in the legislation. With legislation similar to the EU with respect to wheat, Canada also requires gluten to be declared separately. Japan includes wheat, but excludes barley and rye from the requirement, and includes buckwheat, which is not a cereal.

Different countries and regions also vary in the exemptions granted for derived ingredients. In the USA, FALCPA exempts raw agricultural commodities (*i.e.* fresh fruits and vegetables) and highly refined oils derived from any one of the eight major food allergens (milk, eggs, fish, crustaceans and shellfish, tree nuts, peanuts, wheat and soybean) together with any ingredient derived from this highly refined oil (*e.g.* soybean lecithin). The EU has several exemptions,[9] such as that for labelling of highly refined soybean oil following assessment of a dossier demonstrating it does not pose a hazard to individuals with food allergies.[18] These exemptions only partially overlap with those granted in the USA. For example, highly refined peanut oils are exempted from allergen labelling in the USA but not in the EU. In the EU, fish gelatin or isinglass used for fining of alcoholic beverages, such as beer, cider and wine, does not have to be declared; this is not regulated by the US FDA, but by the US Alcohol and Tobacco Tax and Trade Bureau, which has a system of voluntary allergen labelling. In Canada, a different approach has been taken with the focus on protein content, because the protein is the hazard. Highly refined soybean contains negligible amounts of the allergenic protein and, consequently, it does not have to be included on labels as an allergen when used to make products (*e.g.* margarine) that are used subsequently as an ingredient for another food product (*e.g.* cakes and biscuits); instead, it can be labelled as vegetable oil. However, if a culinary oil is less refined, and contains a significant amount of protein, is added to a product, the ingredient must be declared on the end-product label. A similar approach has been adopted in the USA.

In recent years, there have been few (USA) or no (EU) applications for exemptions, presumably as a consequence of waning interest or the scientific and technical challenges associated with such requests. FBOs using exempted raw materials or foods, therefore, need to check regulations to ensure they meet regional specifications, which are still being updated globally. Checks, however, go far beyond the final product labels to include appropriate specifications for ingredients and processing.

7.2.3 Allergen Declarations

Allergenic compound ingredients must be declared in the ingredient list and, although not permitted in the EU, allergens can also be listed separately in a "contains", either in a regulated (such as in the USA) or a voluntary fashion. Separate declarations of allergens (*i.e.* allergen advisories) appear to contribute to recalls, due to discrepancies with ingredient lists.[19] Instead, the EU approach is to require that food allergens are highlighted in bold in the ingredient list (Table 7.4).

Even if general principles seem straightforward, implementation – given the complexity of food supply – leads to practical problems, often specific to product type or legislation. The very need to provide guidance on implementation highlights that these difficulties are perceived as significant by regulatory authorities, *e.g.* US FDA[13,16] and European Commission's DG SANTE[17] amongst others.

FBOs, including distributors and suppliers, must be aware that, whilst labelling errors (*e.g.* incorrect declaration of nutrients) may result in fines or recommendations when identified by authorities, errors that have safety implications for allergic consumers can result in costly recalls. However, even in this respect, there are differences in implementation globally. For example, whilst authorities in Australia, Canada, the USA and most of the EU will recall products as not safe, some EU MS (*e.g.* Germany) are less stringent, on the basis that food products are not specifically for allergic consumers. Thus far, no case has been brought to the European Courts of Justice to challenge this interpretation of Article 14 of Regulation (EC) No 178/2002.

It is also important to note that the design, review and approval of a label is only one element in a complex compliance process. Labels must be compliant but also informative, but this can only be achieved if information on which labelling decisions are based, encompassing the entire supply chain, is exhaustive and correct. Within the EU, prepacked, non-prepacked and catered foods are within the scope of Regulation (EU) No 1169/2011, although implementation with regards to catered foods is delegated to EU MS and hence varies between countries. Thus, in Ireland allergens need to be listed on menus in restaurants whereas in the UK restaurateurs can provide verbal information, although this must be backed up by written information about allergenic ingredients used to prepare dishes. In contrast, there are no such legal requirements in the USA.

Table 7.4 Examples of declaration of allergenic foods in the USA and in the EU.

Labelling options	USA	EU
Option 1: listing of food source	Ingredients: flour (wheat flour, malted barley), whey (milk), eggs, lecithin (soy)	Ingredients: flour (**wheat** flour, malted **barley**), whey (**milk**), **eggs**, lecithin (**soy**)
Option 2: "contains"	Ingredients: flour (wheat flour, malted barley), whey (milk), eggs, lecithin (soy)Contains: wheat, milk, egg, and soy	Not permitted

The goal in extending scope was to help food-allergic consumers avoid problem foods in environments where they are at particular risk of accidental exposure. However, it can be a challenge for catering outlets to achieve this, as many as are small- or micro-sized businesses, with limited technical and regulatory knowledge. A study investigating how food allergen information provision has changed following the change in legislation in Europe showed that, in general, it appears to have improved, but varied amongst the different types of food allergen, with particular issues being identified with regards to milk and milk-allergic consumers being reluctant to ask restaurant staff for alternatives.[20] One of the conclusions of the study was that staff training to improve allergen awareness would improve the experience of allergic consumers when eating out of home.

7.2.4 Voluntary Precautionary Allergen Labelling

Allergens can find their way into food products unintentionally through commingling of agricultural products and the use of common processing lines and shared utensils and equipment. Practices to improve manufacture of different allergen-containing recipes and more stringent cleaning protocols can do much to manage the presence of unintended allergens. However, in some instances, it is difficult to avoid, *e.g.* carry-over of milk from the manufacture of milk and plain chocolates. Because such allergens are not added intentionally, they cannot be included on the ingredient labels and, hence, manufacturers use precautionary allergen labelling (PAL) to warn allergic consumers of their presence.

In the EU, general food law requires that the needs of minority groups, such as food-allergic consumers, be taken into account by FBOs to ensure they have access to foods that are safe to eat. These obligations require manufacturers to manage unintended allergens and warn consumers of potential hazards, if necessary. However, the USA and Canada, amongst many other jurisdictions, require that PAL does not mislead consumers and is not used as a substitute for effective food allergen control measures.[21] There is a general consensus that risk assessment approaches should underpin decisions to apply PAL to ensure such labels are not used reactively to protect FBOs against product liability claims.

One major initiative to develop evidence-based risk assessment approaches has been Voluntary Incidental Trace Allergen Labelling (VITAL™) 2.0 in Australia and New Zealand. VITAL™ has developed a set of reference doses for priority allergens for which labelling is mandatory[22,23] and can be used to identify the need for mitigation strategies (*e.g.* improved cleaning protocols) and, if these are not sufficiently effective, how to use a PAL statement. Other more sophisticated quantitative risk assessment models have also been developed using Monte Carlo simulations, which rely on access to threshold data[24] and have been used to assess how low levels of unintended allergens might cause reactions in the food-allergic population.[25]

At present, despite such efforts, there is no international consensus on reference doses that are considered safe for the majority of food-allergic individuals and on which action levels could be based and trigger the use of PAL. Use of reference doses is not regulated in most jurisdictions and is largely done on a voluntary basis, with the exception of China, which requires some warnings in the case of possible cross-contact, and Argentina[26] and South Africa, where a documented risk assessment has been undertaken to warrant the use of PAL.[27] Japan and Switzerland, which have both limited the use of PAL, have identified threshold levels for unintended allergens. Under Japanese regulations, unintended allergen concentrations greater than 10 mg of allergenic protein per kg of food product have to be included in the ingredients list rather than PAL, which is prohibited. In Switzerland, unintended allergens (apart from gluten) must be in excess of 1000 $mg\,kg^{-1}$ before PAL is applied; PAL for unintended gluten is triggered by levels over 100 $mg\,kg^{-1}$.

A variety of PAL statements have been/are employed, the meanings of which are not always clear in terms of whether a food is safe to eat.[21,28] Examples of PAL range from brief statements such as "may contain....", "not suitable for" to "manufactured in a factory which handles....", "made on equipment also used to process...." and statements such as "We made it in a busy working kitchen so it may also contain traces of nuts and sesame" and "Any allergies? I contain celery. I have been known to hang around near nuts, peanuts, and sesame seeds and I may contain them as well".[21]

Surveys have been carried out in many different parts of the world to determine levels of undeclared allergens, which have found they are often unrelated to the presence of PAL.[29–33] Care must be taken in interpreting data from such surveys, given the poor recoveries achieved by many enzyme-linked immunosorbent assay (ELISA) methods, especially when analysing highly processed food products, meaning allergen concentrations are often under-reported. It is also the case that particulate foods cause sporadic, high levels of unintended allergen in batches of products where the vast majority of packs contain no unintended allergens. Such products need to carry a PAL to warn consumers, *e.g.* a nut fragment might become lodged in food processing equipment and accidentally find its way into only one chocolate bar of many thousands. Nevertheless, even with such caveats, many surveys suggest that such a small proportion of PAL-carrying foods contain unintended allergens, FBOs are being overly cautiously (for legal reasons rather than any other) and applying PAL without adequate risk assessments. As a consequence of the diversity in phraseology used for PAL, and the way it is often applied without real evidence of unintended allergen presence, there is an increasing lack of consumer trust in PAL.[34] It is, therefore, important that FBOs undertake a risk assessment to demonstrate the need for PAL and check local provisions for implementation, rather than assuming the same PAL can be carried in all countries and regions globally.

7.3 Allergen-free Food Regulation and Labelling

The demand for, and availability of, "free-from" foods that have been manufactured using recipes without major allergens (*e.g.* milk, egg, peanuts, tree nuts and gluten) are increasing. This is driven by increasing numbers of individuals with either IgE-mediated food allergies or cell-mediated gluten intolerance syndrome (coeliac disease), together with those individuals avoiding gluten- or lactose-containing foods because of other types of intolerance. It can be difficult to substitute allergens with other ingredients and achieve taste and flavour, appearance and textures comparable to conventional allergen-containing products. This is especially true of wheat-based foods, as the unique viscoelastic nature of gluten, so essential for applications such as bread-making, is difficult to replicate in gluten-free foods. The drive to improve organoleptic qualities for "free-from" foods can also be detrimental to their nutritional quality. Thus, gluten-free foods consistently have higher fat and sugar contents than gluten-containing alternatives and can lead those conforming to a gluten-free diet to gaining weight and a reduced quality of life.[35,36]

There are no specific regulatory frameworks governing allergen-free foods, apart from gluten-free. As a consequence, many jurisdictions implement a "zero tolerance" policy, meaning the allergen about which the free-from claim is made has to be undetectable in the finished product. This can pose problems for FBOs because there are many issues with the reproducibility and comparability of allergen test results[37] and different methods may have different limits of detection. Consequently, analysis undertaken using one kit might yield an "undetectable" result that may not be confirmed by another.

The regulatory framework for "gluten-free" claims is more established, with most regulations around the world based on recommendations laid out in the CODEX Standard 188-1979.[38] This standard defines gluten as "a protein fraction from wheat, rye, barley, oats or their crossbred varieties and derivatives thereof, to which some persons are intolerant and which is insoluble in water and 0.5 M sodium chloride solution". In the EU, regulations pertaining to gluten-free and reduced-gluten foods are also based on the CODEX Standard and the Food Information for Consumers Regulation (EU) No 1169/2011, together with Regulation (EU) No 828/2014.[39] These Regulations replaced the original legislation,[40] but did not change the rules governing the use of "gluten-free" and "very low gluten" statements, but extended them to non-prepacked foods, ensuring labelling of such foods is consistent with other allergen labelling. These regulations also prohibit the use of "gluten-free" or "very low gluten" on infant and follow-on formulae.

EU Regulation No 828/2014 specifies that gluten is a protein fraction originating from cereal species including wheat (including all species of Triticum [such as *Triticum aestivum* (bread wheat), *Triticum durum* (pasta wheat), *Triticum spelta* (spelt) and *Triticum dicoccum* (einkorn)]), rye, barley, oats (*Avena sativa*) or their cross-breed varieties (*e.g.* Triticale, a wheat–rye hybrid) and derivatives thereof. A defined amount of gluten (20 mg of gluten per kg) can occur in foods carrying a "gluten-free" claim, based on CODEX Standard 188-1979. This was set to avoid adverse effects in those with coeliac disease but, without an agreed reference dose, it is not clear whether this is also safe for individuals with IgE-mediated wheat allergy.

The other limit (100 mg gluten per kg), set for "very low gluten", is also based on the CODEX Standard 188-1979; this level would typically elicit gastrointestinal symptoms in the majority of coeliacs.[41] However, foodstuffs labelled "very low gluten" can contain ingredients that have been processed to reduce or remove gluten, as well as non-gluten-containing ingredients, provided total gluten does not exceed 100 $mg\,kg^{-1}$. Regardless of the concentration of gluten in the final product, if a cereal containing gluten is used as an ingredient it must be included in the list of ingredients.[17] The majority of individuals with coeliac disease and IgE-mediated wheat allergy can safely consume oats, which is acknowledged in the CODEX Standard 118-1979. However, this is with the proviso that the oats are free from wheat contamination, which is a significant problem, as contamination of oats can occur along the length of the food chain (*i.e.* in the field or during storage with gluten-containing cereals, such as wheat, barley or rye).[42]

Australia and New Zealand have also modelled their regulations on the CODEX recommendations, but with some significant changes.[8] Gluten is defined only in relation to wheat, spelt, triticale, rye, barley and oats, which can cause coeliac disease or dermatitis herpetiformis. Thus, FSANZ dictates a tiered system for gluten labelling.[8]

1. Gluten-free: a product must have no detectable gluten present, does not contain oats and does not contain gluten-containing cereals that have been malted, or products of such cereals.
2. Low gluten: claim may be made if the gluten content does not exceed 20 mg gluten per 100 g food, *i.e.* 10-fold higher than the level required to achieve a gluten-free claim.
3. Contains gluten or high in gluten: if the gluten content exceeds 20 mg per 100 g food then a contains or high gluten statement can be included on the label.

If a claim (1) and (2) above is made relating to gluten, the average concentration of gluten has to be included, alongside nutrition information for that product in either mg/serving or mg per 100 g. Other types of claims regarding gluten content of foods for special medical purposes are prohibited. FSANZ regulations raise interesting questions about analytical sensitivity. Development of methodologies with increased sensitivity might make it increasingly difficult for manufacturers to make gluten-free claims, due to detection of minute quantities of gluten. However, whether these ever-decreasing limits of detection determine concentrations of gluten that are clinically relevant has to be decided if products are not to be rejected unnecessarily, *i.e.* despite not causing adverse effects in individuals with coeliac disease.

In the USA, gluten-free statements are regulated by an FDA rule, which – much like the EU Regulations – is based on the CODEX recommendations and set at a limit of 20 mg gluten per kg for gluten-free claims.[13] The regulations also cover foods using different phraseologies, including claims such as "no gluten", "free of gluten" or "without gluten" but relate only to packaged foods. The FDA defined gluten as "proteins that naturally occur in a gluten-containing grain and that may cause adverse health effects in persons with celiac disease (*e.g.* prolamins and glutelins)". The FDA also defined gluten-containing grain as any species belonging to the genus Triticum (including wheat), Secale (including rye) and Hordeum (including barley) or any of their crossbred varieties. Thus, one major difference between the USA and the EU is that the former permits oats to be called gluten-free, providing they contain less than 20 $mg\,kg^{-1}$, whilst the latter does not. Refined wheat products, such as starch, which also have less than 20 $mg\,kg^{-1}$ gluten can be called "gluten-free", but have to be labelled as "contains wheat". To avoid the obvious confusion arising from such apparently conflicting labels, FALCPA requires an explanation to be included, *e.g.* refined wheat starch can be marked with an asterisk which refers to the statement, "wheat starch has been processed to allow this food to meet the FDA requirements for gluten-free foods". FALCPA does not define claims such as "low gluten" or "very low gluten" and actively discourages use of such on packaging. If a claim is made, this is reviewed on a case-by-case basis to determine the truthfulness of the statement.

An important aspect in establishing the validity of any claim about gluten content is access to effective, quantitative methodologies. The CODEX Standard 188-1979 indicates the types of methodology, such as ELISA, which should be used to verify whether a food product warrants "gluten-free" or "very low gluten" claims. However, the US FDA does not mandate a specific test for manufacturers to use, recommending only "use of scientifically valid methods in order for the gluten test results obtained to be reliable and consistent". However, when enforcing regulations and determining compliance, the US FDA must use both the Wheat Protein ELISA Kit (Gliadin) from Morinaga Institute of Biological Science, Inc. and RIDASCREEN® Gliadin from R-biopharm.

Japan does not include gluten specifically in food labelling laws but, instead, focuses on wheat as one of seven major food-specific allergenic ingredients that must be declared (Food Labeling Act, Act No. 70, enforced in 2015). This is important for those with coeliac disease, as other cereals containing gluten are not included and, therefore, do not need to be included in the ingredients. There is a defined amount of wheat that can be included in foodstuffs to maintain a gluten-free claim, namely 10 μg soluble protein/g food, equating to 10 ppm. The Japanese Ministry of Health, Labour and Welfare, Department of Food Safety has attempted to harmonise allergen detection methodologies and has stipulated that calibrators must be a mixture of 14 species of wheat extracted 1 in 20 with a standardised extraction buffer. This must be used in conjunction with a standardised reference standard for measurement and two ELISA kits must be used, namely FASTKIT ELISA Ver. II®, FASPEK ELISA® or ALLERGENEYE® ELISA. If results from the two ELISAs differ, manufacturing records must be scrutinised to determine appropriate labelling. Further, if manufacturing records are incomplete or insufficient, a confirmatory polymerase chain reaction test must be undertaken, specifically Wheat Allergen checker [Wheat] produced by Oriental Yeast Co., Ltd.[43]

7.4 Conclusions

The advent of food allergen labelling has undoubtedly improved quality of life for individuals with food allergies and intolerances over the last 15 years or so. This is particularly true for foods where allergenic ingredients are included deliberately in a recipe. However, much uncertainty remains regarding PAL and "free-from" claims. The development of internationally agreed reference doses, which are recognised as being generally safe for most allergic consumers, is essential if the use of PAL is to become transparent and harmonised globally. It should also require that a risk assessment is mandatory and indicated on the label; otherwise the absence of PAL remains unclear to the consumer (*i.e.* is the food safe because no PAL was required or was no risk assessment undertake to determine potential risk?). Such actions could also provide a basis for the making of "free-from" claims for foods/ingredients or components besides gluten and, thus, avoid the situation where foods that are safe for the majority of allergic individuals are rejected on the basis of analytical measurements alone. For companies and food-allergic consumers alike, there is also increasing divergence in regulatory frameworks being implemented globally, which has the potential to reduce the impact of such legislation, particularly with respect to quality of life and food choice. Much has been achieved, but there is still much to do to ensure harmonisation of approaches for allergen labelling of foods.

References

1. E. N. Mills, E. Valovirta, C. Madsen, S. L. Taylor, S. Vieths and E. Anklam, *et al.*, Information provision for allergic consumers–where are we going with food allergen labelling?, *Allergy.*, 2004, **59**(12), 1262–1268.
2. Commission CA, Report of the Nineteenth Session of the Joint FAO/WHO CODEX Alimentarius Commission, 1991.
3. Food allergen database: Guideline for completing allergen data for brand lists (Version 1.0 – 2009), 2009.
4. Commission CA, CODEX general standard for the labelling of prepackaged foods, 1999, p. 1–11.
5. S. L. Taylor, Emerging problems with food allergens, in *Food, Nutrition and Agriculture 26*, ed. J Albert, Food and Agriculture Organization of the United Nations, Rome, 2000, pp. 14–23.

6. M. Shoji, R. Adachi and H. Akiyama, Japanese Food Allergen Labeling Regulation: An Update, *J. AOAC Int.*, 2018, **101**(1), 8–13.
7. European Union Directive 2003/89/EC of the European Parliament and of the Council of 10 November 2003 amending Directive 2000/13/EC as regards indication of the ingredients present in foodstuffs, (*Offi. J. EU*, 2003, **308**, 15–18).
8. Food Standards Code – Standard 1.2.8 – Nutrition Information requirements, 2018. Available from: http://www.foodstandards.gov.au/code/Pages/default.aspx.
9. Regulation (EU) 1169/2011 of the European Parliament and of the Council of 25 October 2011 on the provision of food information to consumers, 2011.
10. PRC NHaFPCo, China National Standard GB7718-2011 General Rule for the Labeling of Pre-packed Foods, 2011.
11. Customs Union Technical Regulations on Food Products Labeling GOST R 51074-2003 Food products. Information for the consumer. General requirements (with Changes N 1, 2), 2012.
12. General standard for the labelling of prepacked foods, 2012. Available from: https://asean.org/wp-content/uploads/2012/10/22PFPWG-Annex-13-ASEAN-Gen-Std-Labelling-Prepackaged-Food-endorsed-22nd-PFPWG.pdf.
13. Guidance for Industry: A Food Labeling Guide (6. Ingredient Lists), 2013.
14. Food Standards Code – Standard 1.2.8 – Nutrition Information requirements, 2019.
15. Canada Go. Food and Drug Regulations C.R.C., c. 870, Division 1 Section B.01.010.1 Canada, 2018.
16. Guidance for Industry: Questions and Answers Regarding Food Allergens, including the Food Allergen Labeling and Consumer Protection Act of 2004 (Edition 4), 2006.
17. Commission Notice 2017/C 428/01 of 13.7.2017 relating to the provision of information on substances or products causing allergies or intolerances as listed in Annex II of Regulation (EU) No 1169/2011 on the provision of food information to consumers, 2017.
18. EFSA Panel on Dietetic Products NaA. Opinion of the Scientific Panel on Dietetic Products, Nutrition and Allergies on a request from the Commission related to a notification from FEDIOL and IMACE on fully refined soybean oil and fat pursuant to Article 6, paragraph 11 of Directive 2000/13/EC- for permanent exemption from labelling, 2007.
19. L. Bucchini, A. Guzzon, R. Poms and H. Senyuva, Analysis and critical comparison of food allergen recalls from the European Union, USA, Canada, Hong Kong, Australia and New Zealand, *Food Addit. Contam., Part A*, 2016, **33**(5), 760–771.
20. J. Barnett, F. M. Begen, M. H. Gowland and J. S. Lucas, Comparing the eating out experiences of consumers seeking to avoid different food allergens, *BMC Public Health*, 2018, **18**(1), 1263.
21. L. M. Jan Mei Soon, "May Contain" allergen statements: facilitating or frustrating consumers?, *J. Consum. Policy*, 2017, **40**(4), 447–472.

22. S. L. Taylor, J. L. Baumert, A. G. Kruizinga, B. C. Remington, R. W. Crevel and S. Brooke-Taylor, *et al.*, Establishment of reference doses for residues of allergenic foods: report of the VITAL expert panel, *Food Chem. Toxicol.*, 2014, **63**, 9–17.
23. K. J. Allen, B. Remington, J. L. Baumert, R. W. Crevel, G. F. Houben, S. Brooke-Taylor, A. G. Kruizinga and S. L. Taylor, Allergen reference doses for precautionary labeling (VITAL 2.0): clinical implications, *J. Allergy Clin. Immunol.*, 2014, **133**(1), 156–164.
24. M. Q. Spanjersberg, A. G. Kruizinga, M. A. Rennen and G. F. Houben, Risk assessment and food allergy: the probabilistic model applied to allergens, *Food Chem. Toxicol.*, 2007, **45**(1), 49–54.
25. M. Q. Spanjersberg, A. C. Knulst, A. G. Kruizinga, G. Van Duijn and G. F. Houben, Concentrations of undeclared allergens in food products can reach levels that are relevant for public health, *Food Addit. Contam., Part A*, 2010, **27**(2), 169–174.
26. M. C. Lopez, Food Allergen Labeling: A Latin American Approach, *J. AOAC Int.*, 2018, **101**(1), 14–16.
27. Regulations relating to the labelling and advertising of foodstuffs., 2010.
28. G. A. Zurzolo, M. L. Mathai, J. J. Koplin and K. J. Allen, Hidden allergens in foods and implications for labelling and clinical care of food allergic patients, *Curr. Allergy Asthma Rep.*, 2012, **12**(4), 292–296.
29. S. E. Khuda, G. M. Sharma, D. Gaines, A. B. Do, M. Pereira and M. Chang, *et al.*, Survey of undeclared egg allergen levels in the most frequently recalled food types (including products bearing precautionary labelling), *Food Addit. Contam., Part A*, 2016, **33**(8), 1265–1273.
30. S. E. Khuda, G. M. Sharma, D. Gaines, A. B. Do, M. Pereira and M. Chang, *et al.*, Survey of undeclared soy allergen levels in the most frequently recalled food categories with or without precautionary labelling, *Food Addit. Contam., Part A*, 2016, **33**(8), 1274–1282.
31. L. Decastelli, S. Gallina, D. Manila Bianchi, S. Fragassi and P. Restani, Undeclared allergenic ingredients in foods from animal origin: survey of an Italian region's food market, 2007-2009, *Food Addit. Contam., Part B*, 2012, **5**(3), 160–164.
32. M. Pele, M. Brohee, E. Anklam and A. J. Van Hengel, Peanut and hazelnut traces in cookies and chocolates: relationship between analytical results and declaration of food allergens on product labels, *Food Addit. Contam.*, 2007, **24**(12), 1334–1344.
33. O. N. Robertson, J. O. Hourihane, B. C. Remington, J. L. Baumert and S. L. Taylor, Survey of peanut levels in selected Irish food products bearing peanut allergen advisory labels, *Food Addit. Contam., Part A*, 2013, **30**(9), 1467–1472.
34. P. J. Turner, K. J. Allen, S. Mehr and D. E. Campbell, Knowledge, practice, and views on precautionary allergen labeling for the management of patients with IgE-mediated food allergy–a survey of Australasian and UK health care professionals, *J. Allergy Clin. Immunol. Pract.*, 2016, **4**(1), 165–167.

35. K. A. Bascunan, M. C. Vespa and M. Araya, Celiac disease: understanding the gluten-free diet, *Eur. J. Nutr.*, 2017, **56**(2), 449–459.
36. L. Norsa, R. Shamir, N. Zevit, E. Verduci, C. Hartman and D. Ghisleni, *et al.*, Cardiovascular disease risk factor profiles in children with celiac disease on gluten-free diets, *World J. Gastroenterol.*, 2013, **19**(34), 5658–5664.
37. P. E. Johnson, N. M. Rigby, J. R. Dainty, A. R. Mackie, U. U. Immer and A. Rogers, *et al.*, A multi-laboratory evaluation of a clinically-validated incurred quality control material for analysis of allergens in food, *Food Chem.*, 2014, **148**, 30–36.
38. Commission CA, Standard for Foods for Special Dietary use for Persons Intolerant to Gluten, 2008.
39. Commission Implementing Regulation (EU) No 828/2014 on the requirements for the provision of information to consumers on the absence or reduced presence of gluten in food, 2014.
40. European Union Commission Regulation (EU) No 41/2009 concerning the composition and labelling of foodstuffs suitable for people intolerant to gluten, 2009.
41. A. K. Akobeng and A. G. Thomas, Systematic review: tolerable amount of gluten for people with coeliac disease, *Aliment. Pharmacol. Ther.*, 2008, **27**(11), 1044–1052.
42. T. B. Koerner, C. Cleroux, C. Poirier, I. Cantin, A. Alimkulov and H. Elamparo, Gluten contamination in the Canadian commercial oat supply, *Food Addit. Contam., Part A*, 2011, **28**(6), 705–710.
43. Japan Go. Ministry of Health, Labor and Welfare of Japan, Department of Food Safety Notification No.1106001, Guidelines for Evaluating Test Methods for Food Products Containing Allergic Substances, Tokyo, Japan, 2002.

CHAPTER 8

The Impact of "Labelling" on the Beliefs, Attitudes and Behaviours of Consumers with Food Allergy: A Multilevel Perspective

AUDREY DUNN GALVIN[a,b]

[a] School of Applied Psychology, University College Cork, Ireland;
[b] Paediatrics and Child Infectious Diseases, First Moscow State Medical University, Russia
Email: A.DunnGalvin@ucc.ie

8.1 Introduction

Although systematic review and meta-analyses have reported the pooled lifetime prevalence of self-reported food allergy as 17.3% (95% CI: 17.0–17.6),[1] food allergy is a major public health concern affecting up to 20 million Europeans (*ca.* 2% of adults and 5–8% of under-16s) and is associated with high costs for public health services,[2,3] if not managed appropriately. Avoidance of trigger food(s) is central to effective self-management, but unintentional consumption is common, causing frequent reactions, some of which may be life-threatening.[4–10] Admission rates for anaphylaxis increased approximately three-fold between 2005 and 2012.[8–10] Furthermore, studies

Food Chemistry, Function and Analysis No. 22
Health Claims and Food Labelling
Edited by Siân Astley

Published by the Royal Society of Chemistry, www.rsc.org

have shown a strong adverse impact of food allergy on health-related quality of life (HRQL) across the life course.[11–15]

Given that, currently, there is no curative therapy for food allergy in routine use, clinical management is, by necessity, focused on avoidance of allergens and effective rescue therapy.[16,17] Strict avoidance ostensibly applies only to food but, because food is such an integral part of everyday life, the impact extends far beyond mealtimes. Food-allergic individuals – and their families – can experience food allergy-related fears 24 hours a day, 7 days a week, whilst on holiday, at parties and in school, using public transportation, and even at home. In food allergy, the risk of reaction is ever present but the when, how and what lead to uncertainty, fear and anxiety.

Confidence in self-management is associated with less stress and anxiety.[18] However, current food labelling practices, despite existing pan-European legislation (Annex II of the EU Food Information for Consumers Regulation No 1169/2011 and Commission Delegated Regulation (EU) No 78/2014 amending Annex II to Regulation (EU) No 1169/2011), present major challenges for consumers, regardless of whether they have a food allergy or are care-givers for someone who is affected. Successful management necessitates access to accurate and comprehensible information about the presence of allergens in food and beverage products whether prepackaged or purchased in retail (*e.g.* hotels, restaurants and other catering outlets, HoReCa). However, the proliferation of precautionary allergen labelling (PAL),[19–21] and lack of both analytical and commercial consensus around their use, means such warnings or, indeed, their absence does not communicate clearly to consumers the risk of exposure.[16]

The issue of communicating risk, due to the presence of specific allergens in a food product, is of particular concern. In Europe, regulations identify 14 foods as priority allergens, the use of which as ingredients or processing aids must be declared on the label or made available to the consumer.[22] However, for unintended allergen presence (*i.e.*, not an ingredient or processing aid), which occurs primarily through manufacturing cross-contact, there are no well-defined, consistent rules governing declaration.[16] This has led to confusion and uncertainty amongst regulators and food business operators (FBOs) and, most importantly, consumers regarding the meaning of PAL statements (*e.g.* may contain nuts). Risk and anxiety are, therefore, increased while confidence, quality of life and food choice are reduced, and a harmonised approach is needed urgently.

Harmonised approaches must be based on a multilevel intervention perspective. In the past, generally, research efforts concentrated on single-level analysis, but these ignore the environment in which individuals operate or how changes in (societal) organisations or policies impact individuals and *vice versa*.[23,24] In the health context, a central goal is to create environments that support healthiness, which is where metaphors, such as "upstream" and "downstream" interventions, arose. However, debate about the merits of upstream (causes of risky behaviour) *versus* downstream approaches (high-quality hospital emergency care for fractures), and whether funding should be at the levels of individuals or populations, ignore the importance of interaction between the two. To better understand the complex nature of systems, Kozlowski and Klein[25] proposed that researchers utilise "approaches that are more integrative, that cut across multiple levels, and that seek to understand phenomena from a combination of perspectives" (page 77). Multilevel approaches are based on a range of systems theories and models[26,27] that, although they may differ in terminology, have at their core the importance of taking into account interactions amongst structures within and across levels,[26] *i.e.*, in the case of food allergy, individuals, FBOs, regulators and policy-makers, and healthcare. This multilevel systems approach holds the promise of representing real-life situations more accurately, leading to innovative solutions to complex challenges.

For the purposes of food allergy labelling, two terms derived from Bronfenbrenners's bio-ecological systems model[27] were employed to examine the impact of PAL on the beliefs, attitudes and behaviours of consumers with food allergy. Firstly, the microsystem, which is the level closest to the individual and contains structures with which the individual has direct contact,[27] includes inter- and intra-personal and contextual influences (*e.g.* family, community, schools, and primary and secondary healthcare) as well as the developmental process (*e.g.* a child being able to cope well with disappointment at not being chosen for the school football team). Secondly, the macrosystem, which is the outermost level in an individual's environment and comprised of culture, norms, policy and regulations.[27] The effects of principles defined by the macrosystem cascade through interactions in all other layers and, thus, by integrating findings at this level, it is possible to suggest more harmonised approaches for labelling that might help to resolve current adverse impacts.

8.2 The Individual (Micro) System

Using and interpreting PAL is a major source of stress for individuals and families living with food allergy. PAL also acts as a barrier to effective management because the lack of clarity around meaning that, whether present or absent, is one of the factors responsible for influencing an individual's perceptions of risk and associated health beliefs around food allergy.

PAL is meant to inform consumers about a demonstrable and significant risk to an individual with a specific food allergy (*e.g.* peanuts). Currently, however, voluntary practices have resulted in huge variations not only in the wording used for these warnings (*e.g.* may contain, may contain traces of, made in the same factory as), but also whether or not they are used.[28,29] A product might be labelled correctly with PAL on the basis of an adequate risk assessment, or to minimise liability without a risk assessment or appropriate allergen management.[28–31] Alternatively, a product without PAL might be completely safe or the unintended presence of an allergen might not have been considered, meaning it is potentially unsafe.[28–31]

Food allergies have psychosocial behavioural components and, therefore, risk perception theories can understand the impact of labelling practices on cognition (understanding), emotions and behaviours (response/choices) of food-allergic individuals. Risk perception is a subjective judgement that individuals make about the characteristics and severity of a risk (*e.g.* whether or not to fly or wear a seatbelt) and the likelihood of negative occurrences, such as injury, illness, disease, prosecution and death.[32] Risk perception does not develop or take place in a vacuum, but in the environmental context of everyday life,[32,33] and has two primary dimensions, namely cognitive (*i.e.*, how much people know about and understand about the risk) and emotional (*i.e.*, how they feel about the risk). Risk perceptions are important precursors to health-related behaviours and management behaviours, including coping with (managing) or preventing risks.[34–36]

One theoretical framework that incorporates both cognitive and emotional dimensions of risk perceptions is the psychometric paradigm developed by Slovic.[37] According to this psychometric paradigm, individuals judge "riskiness" of a hazard based on a range of (perceived) risk characteristics that include whether the severity of the risk can be controlled, whether the risk makes people feel dread and is experienced in unequal ways (*e.g.* experiencing an allergic reaction *versus* watching someone having an allergic reaction), whether the risk is increasing or exposure to the risk is involuntary, whether the risk affects an individual personally, whether people know if they are exposed to the risk, and if effects are immediate or the risk is unknown to science.[37] Covello[38] proposed that similar factors influence people's risk perceptions, including voluntariness, controllability, understanding, uncertainty, dread, trust in institutions, reversibility and personal stakes, which are used to inform risk communication strategies. "Mental noise" or stress arising from these factors, however, reduces the ability of individuals to process risk-related information effectively. In particularly, uncertainty is a major obstacle in constructive decision-making for all age groups[33,37,38] and is also a central theme in living with food allergy.[15,16,28]

Many consumers believe erroneously that PAL is regulated and mandatory. This perception drives beliefs not only that the presence of PAL has a clear meaning, but also that absence implies safety. In turn, this belief drives the misperception that variations in labelling statements reflect a hierarchy of risk, *e.g.* "may contain" – in itself permissive (may) rather than indicative (might) – indicates a higher risk than "may contain traces".[16] In reality, there is no relationship between the risk of contamination and the wording employed.[39,40] Taking risk perception theory into account, this hierarchy of risk might be a search for control and a reduction in uncertainty and dread. Monks *et al.* (2010) found that food-allergic teenagers avoided PAL labels, depending on the wording, with "may contain" perceived to denote greater risk than "made in the same factory as".[41] This finding was confirmed in a multinational survey conducted by the Patient Organisation Committee of the European Academy of Allergy and Clinical Immunology and the international Food Allergy and Anaphylaxis Alliance (iFAAA, personal communication, Sabine Schnadt). Similar misperceptions about the wording of PAL were found in a survey of 239 UK-based health professionals (136 doctors, 47 dieticians and 47 clinical nurse specialists),[42] with many commenting on consumption of foods with statements that might be interpreted as conveying a lower risk (*e.g.* "made in an environment" was regarded wrongly as less significant than "may contain" statements).[42]

Another misperception is that PAL refers to an unintended presence of allergen that is present consistently or present only in trace amounts. So, when a product does not cause a reaction, some consumers feel the product is safe or they can tolerate the amount of allergen that is present.[16] This may lead to unintended risks and, when a reaction does occur, to a loss of trust and PAL credibility as well as the science underpinning it, creating anxiety, uncertainty and dread. Individuals do not know whether they will be exposed to the risk, and the risk is seen as involuntary and unknown to science, meaning trust in the institutions responsible for food safety is also impacted negatively.

The proliferation of PAL, together with its use on unexpected products (*e.g.* may contain nuts on a nut-based product), has led to cynicism amongst consumers and deliberate risky behaviours. In effect, consumers have a dilemma: observe the warning and experience decreased choice or ignore the warning and risk an allergic reaction. This is evidenced by the relationship between the extent to which PAL is used by FBOs and observed by food-allergic consumers.[16] We know that many consumers with food allergy take deliberate (calculated) risks.[39–41,43,44] Sampson *et al.* (2006) reported that risk-taking behaviour was associated with levels of perceived risks; more than half of respondents had knowingly eaten a small amount of allergen-containing foods because of previous experience (without a reaction), wish to eat the food, reluctance to ask questions or peer pressure, which has been confirmed by more recent studies.[45]

Consumers with food allergies might also respond to risks by being very selective, avoiding enjoyable (sociable) activities like dining out, travelling, shopping, *etc.*[46] Research on HRQL found some of the strongest, and most adverse, impacts of food allergy in relation to social and dietary restrictions across all age groups and care-givers, including fear of new foods, new people and new places,[12–15] which is understandable when the media reports devastating consequences (*e.g.* the death of Natasha Ednan-Laperouse in 2016). However, restrictions on diet not only affect children's nutrition, but also school life and extracurricular activities. In Italy, 20% of food-allergic children have never attended a birthday party and 40% of children reported having a monotonous diet.[47] Reactions that occur in schools are, in part, due to uncertainty around food safety amongst school authorities, meaning the potential risks are managed inappropriately by excluding or restricting children from participation in certain activities or setting them apart from others during mealtimes, which creates an environment giving rise to feelings of difference, stigmatisation and, sometimes, bullying by peers.[12–15,48]

The indirect impact of PAL may be long term, because developmental processes programme health and well-being over the life course,[49,50] *i.e.*, lifelong patterns of behaviour (*e.g.* participation in social events) are established in childhood and determine current and future health outcomes.[49,50] While any chronic condition (*e.g.* asthma, diabetes, epilepsy) can potentially affect developmental processes, the reverse is also true, *i.e.*, physiological, environmental and psychosocial factors can impact the disease. Chronic diseases associated with dysregulation of inflammation (*e.g.* eczema, psoriasis), including food allergy, are particularly susceptible to modulation by stress and emotion.[50] Significant fear-eliciting experiences in childhood due to inappropriate risk management can disrupt development of stress regulation as well as learning, memory and social behaviour. Anger, embarrassment, feelings of frustration, dread, fear and anxiety are just some of the psychological outcomes of day-to-day management of food allergy without the additional impact of deliberate exclusion.[49] Uncertainty around reactions, due in part to PAL, has been shown to take root in young children with poor outcomes seen later in developmental pathways. Food-allergic adolescents (12–25 years) have higher incidences of internalising problems (*i.e.*, anxiety, depression, relational difficulties, lower HRQL), with greater impact on psychological problems in everyday life than non-allergic peers. Young people are also at highest risk of anaphylaxis, arising from risky behaviours.[40,45,47] Sharing food and social eating experiences have important social value and uncertainty about PAL does not help attempts to negotiate the tricky balance amongst safety, (social) identity and autonomy in life.

8.3 The Society (Macro) System

In order to better understand and address the challenge of PAL and related practice, the wider manufacturing, scientific and institutional factors that give rise to current use need to be examined. Situations generating risks, and subsequent emotions, such as fear, dread and anxiety, create an environment where individuals are more likely to focus on negative messages. Risk perception theories underline the importance of perceived trust in the communicator on perceptions and responses/reactions, highlighting the importance of trust determination factors, such as competence and expertise, honesty and openness.[32–38] Trust, as a concept, has been studied extensively in relation to consumer perceptions of food safety and food risk management in general. Trust in different actors and institutions responsible for guaranteeing and controlling food safety, as well as information provided by different sources about food safety or food-related risks, is important for consumer confidence.[33,34]

The food industry recognises that PAL has spread to such an extent that, in certain product categories, the majority of products bear such warnings. A 2006 US survey revealed that 17% of retail food products had PAL warnings with more than 40% of chocolates, biscuits and baking mixes carrying PAL.[51]

Similarly, 65% of Australian retail food products from selected categories have PAL.[52] Several analytical surveys have documented that only a small percentage of products with peanut PAL contain detectable allergen residues and some without advisory labels contain similar levels.[53–56] Furthermore, the percentage of PAL products that contain detectable peanut was less than 10%[53–56] with similar results for other food allergens.[53–56] In other words, PAL warnings are meaningless in terms of exposure to risk. Thus, risk perception factors, such as lack of understanding, uncertainty, dread and control, which drive food-allergic consumer choices, might also be driving increasing and inconsistent use of PAL, particularly amongst small and medium enterprises (SMEs), rather than risk assessments that have revealed low or negligible cross-contact. This is a major challenge as SMEs represent the majority of FBOs in Europe (99.1% in the EU) and contribute a significant proportion of the food supply in most Member States (49.3% of turnover in the EU).[57] SMEs also represent a large proportion of ingredients suppliers to companies as well as producing retailers' own-brand products.[57]

Analytical uncertainties also exist and may contribute to decisions leading to more widespread use of PAL.[28,53–56] Enzyme-linked immunosorbent assays (ELISAs) are the favoured analytical approach for FBOs, because they are relatively easy to use and detect protein/allergen residues in a highly specific and sensitive manner.[28,53–55] Although a consensus exists in terms of validation of quantitative ELISA kits,[53–55] most commercial kits have not been validated, and no consensus yet exists for validation of lateral flow devices, an easy-to-use form of ELISA that is very popular within the industry. In part, this is because reference standards do not exist for the majority of allergenic foods, a deficiency that inhibits method validation.[28] Another factor is the potential for cross-contact with particulates containing allergenic components. Particulates are common in certain segments of the food industry (*e.g.* baking) and the presence of particulates (a piece of peanut compared with peanut flour) creates sampling problems that impact confidence, controllability and trust in analytical results: an allergen might not be detected if the sample collected does not contain the particle. This issue may lead FBOs to use PAL, even in the absence of positive analytical results identifying the presence of an allergen. Thus, although regulatory authorities advise that PAL should only be applied following a thorough risk assessment, its use might stem from a lack of guidance on risk assessment as a decision-making tool for labelling.

There is no specific European legal provision (Directive or Regulation) that determines when PAL should be applied. However, EU Regulation 178/2002 (https://eur-lex.europa.eu/legal-content/EN/ALL/?Uri=ncelex%3A32002R0178) on the general requirements of food law, states that "food must not be placed on the market if it is injurious to health". However, the amount of an allergen – internationally agreed quantitative reference doses – that is "injurious to health" is yet another area of uncertainty, which regulators cite for the lack of action in this respect across Europe.

8.4 What Can Help From a Multilevel Perspective?

The credibility of PAL, and thereby its value as a risk management tool, will only be restored if the warnings communicate a clear message about risk to food-allergic consumers. This can only be achieved through adoption of a common standard across the industry (including ingredient suppliers), clearly explained and communicated to consumers, and preferably endorsed by food safety authorities. One of the strong drivers for a "safety net" approach is an emphasis on hazard control, rather than risk assessment, in both regulation and manufacture of products. Internationally agreed reference doses need to be derived from allergen threshold doses to determine action levels below which PAL need not used; only a very few food-allergic individuals react to any amount of food allergens. Thus, threshold doses should inform decisions by regulators handling incidences of unintended exposure.

Quantitative risk assessment (QRA) is emerging as an approach that can be used to predict potential risks associated with allergen residues in foods,[53–55] such as the expected number of reactions in PAL products.[55] QRA requires inputs from threshold dose distributions for specific food allergens, consumption levels across the population, and analytical data describing the product. In this case, manufacturers will be required to assess the amount of each allergen in their product. If any of these will cause a reaction in at least one in 100 allergic consumers, a "may contain" statement will be mandatory. While guidance on thresholds has yet to emerge, threshold dose distributions do exist for 11 priority foods.[55] We now know, for example, the thresholds at which 95% and 99% of the population can eat foods containing peanuts before having a reaction.[55] Research is also ongoing into the stability of individuals' thresholds, *i.e.* levels of exposure that individuals can reliably not react if exposed. These doses are sufficient for QRA and have been used as the basis for an approach to PAL, specifically VITAL (Voluntary Incidental Trace Allergen Labelling, allergenbureau.net/vital) developed by the Allergen Bureau of Australia and New Zealand.

To date, the use of QRA by FBOs is minimal and must be expanded, as a means to make PAL more meaningful to food-allergic consumers. Guidance from regulatory authorities in this regard could speed up adoption of this approach. If reference doses can be established, and used to guide labelling decisions, the presence of a specific advisory warning label (AWL) could be related to defined levels of risk, as with VITAL. Previous research has shown that practical interventions can reduce uncertainty and improve quality of life,[58] allowing families to feel more in control of their lives, diet and nutrition, and health. For the same reasons, harmonised actions around food allergens would reduce uncertainty and improve quality of life. In terms of acceptability, research has shown that a lack of meaningful communication[38] amongst stakeholders is a major barrier to implementation. Thus, helping healthcare professionals and allergic consumers understand QRA and AWL is key to successful adoption.

Once there are internationally agreed reference doses, regulators will need to consider how best to communicate QRA and AWL. Consideration of the most appropriate phrases is required at both levels, *i.e.* when contamination is above AND below the threshold specific to that priority food. Communications need to be consistent and meaningful, *i.e.* that there is minimal (not zero) risk of a reaction and, should there be a reaction, it is unlikely to be severe. Research, under the aegis of iFAAM (Integrated Approaches to Food Allergen and Allergy Risk Management, Project ID: 312147, research.bmh.manchester.ac.uk/iFAAM) has shown that consumers want to know what processes have taken place to result in a warning label on a food product.[59] A combination of statements about unintended allergen presence and risk assessment and a risk assessment symbol was the preferred option, despite the presumption that this might be more confusing for consumers. Similar preferences were also found in previous research.[60]

A multilevel systems approach holds more promise in accurately representing real-life situations and, thus, with proper research design and methods, facilitating effective and efficient resolutions for system-wide challenges like food allergy. When expert knowledge – which includes food-allergic individuals as well as healthcare, the food industry, research and policy-makers and regulators – is integrated effectively, and focused on solving problems, it can yield significant innovation. Central themes in food allergy relate to the overuse and/or misuse of PAL and uncertainty, dread, mistrust and lack of control. The resulting noise increases anxiety and frustration and leads to unnecessary avoidance or risky behaviours, with direct and indirect effects on emotional adjustment, food choice, nutrition and other related health risks, social interaction, development, confidence and quality of life. Simply put, decisions about product labelling impact significantly beliefs, attitudes and behaviours of consumers with food allergy, and consumers need to be able to trust and use allergy warning labels in a meaningful and consistent way. Where they are unavoidable and necessary, they should be regulated with prescribed wording, much like other health claims. Where appropriate, there should be a clear indication that a product has undergone risk assessment and does not pose a risk for food-allergic consumers. When an AWL is required, it must be based on internationally agreed threshold levels, derived from scientific evidence that will protect consumers from severe allergic reactions. Most importantly, a communication strategy about the concept and meaning of AWL and the related science and regulations must be developed for food-allergic individuals and their families as well as stakeholders including healthcare professionals (*i.e.* clinicians and dietitians).

References

1. (a) B. I. Nwaru, L. Hickstein, S. S. Panesar, G. Roberts, A. Muraro and A. Sheikh, On behalf of The EAACI Food Allergy & Anaphylaxis Guidelines Group, Prevalence of common food allergies in Europe: a

systematic review and meta-analysis, *Allergy*, 2014, **69**, 992–1007; (b) S. Prescott and K. J. Allen, Food allergy: riding the second wave of the allergy epidemic, *Pediatr. Allergy Immunol.*, 2011, **22**, 155–160.
2. A. W. Burks, M. Tang and S. Sicherer, *et al.*, ICON: food allergy, *J. Allergy Clin. Immunol.*, 2012, **129**, 906–920.
3. R. S. Gupta, E. E. Springston and M. R. Warrier, *et al.*, The prevalence, severity, and distribution of childhood food allergy in the United States, *Pediatrics*, 2011, **128**, e9–e17.
4. W. K. Liew, E. Williamson and M. L. Tang, Anaphylaxis fatalities and admissions in Australia, *J. Allergy Clin. Immunol.*, 2009, **123**, 434–442.
5. R. J. Mullins, K. B. Dear and M. L. Tang, Time trends in Australian hospital anaphylaxis admissions in 1998-1999 to 2011-2012, *J. Allergy Clin. Immunol.*, 2015, **136**, 367–375.
6. P. J. Turner, J. L. Baumert, K. Beyer, R. J. Boyle, C. H. Chan and A. T. Clark, *et al.*, Can we identify patients at risk of life-threatening allergic reactions to food? *Allergy*, 2016, **71**(9), 1241–1255.
7. A. Versluis, A. C. Knulst, A. G. Kruizinga, A. Michelsen, G. F. Houben, J. L. Baumert and H. van Os-Medendorp, Frequency, severity and causes of unexpected allergic reactions to food: a systematic literature review, *Clin. Exp. Allergy*, 2015, **45**, 347–367.
8. P. J. Turner, M. H. Gowland and V. Sharma, *et al.*, Increase in anaphylaxis-related hospitalizations but no increase in fatalities: An analysis of United Kingdom national anaphylaxis data, 1992-2012, *J. Allergy Clin. Immunol.*, 2015, **135**(4), 956–963.
9. A. M. Branum and S. L. Lukacs, Food allergy among children in the United States, *Pediatrics*, 2009, **124**, 1549–1555.
10. S. H. Sicherer, J. A. Forman and S. A. Noone, Use assessment of self-administered epinephrine among food-allergic children and pediatricians, *Pediatrics*, 2000, **105**, 359–362.
11. A. J. Cummings, *et al.*, The psychosocial impact of food allergy and food hypersensitivity in children, adolescents and their families: a review, *Allergy*, 2010, **65**(8), 933–945.
12. G. D. Pinczower, N. A. Bertalli and N. Bussmann, *et al.*, The effect of provision of an adrenaline autoinjector on quality of life in children with food allergy, *J. Allergy Clin. Immunol.*, 2013, **131**, 238–240.e1.
13. B. M. J. Flokstra-de Blok, *et al.*, Health-related quality of life of food allergic patients: comparison with the general population and other diseases, *Allergy*, 2010, **65**, 238–244.
14. N. J. Goossens, B. M. Flokstra-de Blok, G. N. van der Meulen, M. H. Arnlind, R. Asero and L. Barreales, *et al.*, Health-related quality of life in food-allergic adults from eight European countries, *Ann. Allergy, Asthma, Immunol.*, 2014, **113**, 63–68e1.
15. A. DunnGalvin, E. Koman, E. Raver, H. Frome, M. Adams and A. Keena, *et al.*, An examination of the Food Allergy Quality of Life Questionnaire performance in a Countrywide American Sample of

Children: Cross cultural differences in age and impact in the US and Europe, *J. Allergy Clin. Immunol. Pract.*, 2017, **5**, 363–368, e2.
16. A. DunnGalvin, C.-H. Chan, R. Crevel, K. Grimshaw, R. Poms and S. Schnadt, *et al.*, Precautionary allergen labelling: perspectives from key stakeholder groups, *Allergy*, 2015, **70**, 1039–1051.
17. A. Muraro, T. Werfel, K. Hoffmann-Sommergruber, G. Roberts, K. Beyer and C. Bindslev-Jensen, *et al.*, on behalf of the EAACI Food Allergy and Anaphylaxis Guidelines Group, EAACI Food Allergy and Anaphylaxis Guidelines. Diagnosis and management of food allergy, *Allergy*, 2014, **69**, 1008–1025.
18. J. Clarke, J. Proudfoot, M.-R. Birch, A. E. Whitton, G. Parker, V. Manicavasagar, V. Harrison, H. Christensen and D. Hadzi-Pavlovic, Effects of mental health self-efficacy on outcomes of a mobile phone and web intervention for mild-to-moderate depression, anxiety and stress: secondary analysis of a randomised controlled trial, *BMC Psychiatry*, 2014, **14**, 272.
19. S. Birot and P. B. Brockhoff, Modelling allergenic risk, DTU Compute PHD-2016, 2017 Kgs. Lyngby, Technical University of Denmark (DTU), No. 441.
20. M. J. Walker, D. T. Burns, C. T. Elliott, M. H. Gowland and E. N. C. Mills, Flawed food allergen analysis–health and supply chain risks and a proposed framework to address urgent analytical needs, *Analyst*, 2016, **141**, 24–35.
21. W. W. Blom and A. D. Michelsen-Huisman, Accidental food allergy reactions: products and undeclared ingredients, *J. Allergy Clin. Immunol.*, 2018, **142**, 865–875.
22. The EU Food Information for Consumers Regulation No. 1169/2011.
23. E. Blas and A. S. Kurup, *Equity, Social Determinants and Public Health Programmes*, World Health Organization, Geneva, 2010.
24. *Social Determinants Approaches to Public Health: From Concept to Practice*, ed. E. Blas *et al.*, World Health Organization, 2011, ISBN 978 92 4 156413 7, NLM classification: WA.
25. J. Turner, K. F. Petrunin and J. Allen, Developing Multilevel Models for Research, in *Handbook of Research on Scholarly Publishing and Research Methods*, ed. V. X. Wang, Florida Atlantic University, USA, 2015, pp. 463–489.
26. N. Luhmann, *Introduction to Systems Theory*, Polity Press, Cambridge, 2013.
27. U. Bronfenbrenner, *Making Human Beings Human*, Sage Publications, Thousand Oaks, 2005, p. xii.
28. A. DunnGalvin, C.-H. Chan, R. Crevel, K. Grimshaw, R. Poms, S. Schnadt, S. L. Taylor, P. Turner, K. J. Allen, M. Austin, A. Baka, J. L. Baumert, S. Baumgartner, K. Beyer, L. Bucchini, M. Fernández-Rivas, K. Grinter, G. F. Houben, J. Hourihane, F. Kenna, A. G. Kruizinga, G. Lack, C. B. Madsen, E. N. C. Mills, N. G. Papadopoulos, A. Alldrick, L. Regent, R. Sherlock, J.-M. Wal and

G. Roberts, Precautionary allergen labelling: perspectives from key stakeholder groups, *Allergy*, 2015, **70**, 1039–1051.
29. M. Worm, A. Moneret-Vautrin, K. Scherer, R. Lang, M. Fernandez-Rivas, V. Cardona, M. L. Kowalski, M. Jutel, I. Poziomkowska-Gesicka, N. G. Papadopoulos, K. Beyer, T. Mustakov, G. Christoff, M. B. Bilò, A. Muraro, J. O. Hourihane and L. B. Grabenhenrich, First European data from the network of severe allergic reactions (NORA), *Allergy*, 2014, **69**, 1397–1404.
30. M. Abbott, S. Hayward, W. Ross, S. B. Godefroy, F. Ulberth, A. J. van Hengel, J. Roberts, H. Akiyama, B. Popping, J. M. Yeung, P. Wehling, S. L. Taylor, R. E. Poms and P. Delahaut, Validation procedures for quantitative food allergen ELISA methods: community guidance and best practices, *J. AOAC Int.*, 2010, **93**, 442–450.
31. S. L. Hefle, T. J. Furlong, L. Niemann, H. Lemon-Mule, S. Sicherer and S. L. Taylor, Consumer attitudes and risks associated with packaged foods having advisory labeling regarding the presence of peanuts, *J. Allergy Clin. Immunol.*, 2007, **120**, 171–176.
32. P. Hye-Jin and T. Hove, Risk Perceptions and Risk Characteristics, in *Oxford Research Encyclopedia, Communication*, Oxford University Press, 2017.
33. D. Kahneman, P. Slovic and A. Tversky, *Judgment under Uncertainty: Heuristics and Biases*, Cambridge University Press, Cambridge, UK, 1982.
34. J. De Jonge, L. Frewer, H. Van Trijp, R. J. Renes, W. De Wit and J. Timmers, Monitoring consumer confidence in food safety: An exploratory study, *Br. Food J.*, 2004, **106**, 837–849.
35. (a) J. Hansen, L. Holm, L. Frewer, P. Robinson and P. Sandøe, Beyond the knowledge deficit: recent research into lay and expert attitudes to food risks, *Appetite*, 2003, **41**, 111–121; (b) S. Henson and J. Northen, *J. Agric. Econ.*, 2000, **51**(1), 90–105.
36. S. Hornig Priest, H. Bonfadelli and M. Rusanen, The "trust gap" hypothesis: Predicting support for biotechnology across national cultures as a function of trust in actors, *Risk Anal.*, 2003, **23**(4), 751–766.
37. *The Feeling of Risk: New Perspectives on Risk Perception*, ed. P. Slovic, Earthscan, Sterling, VA, 2010.
38. V. T. Covello, D. Winterfeldt and P. Slovic, Communicating scientific information about health and environmental risks: problems and opportunities from a social and behavioral perspective, in *Uncertainty in Risk Assessment, Risk Management, and Decision Making*, ed. Covello *et al.*, Plenum Press, New York, 1987.
39. A. DunnGalvin, A. Gaffney and J. O. B. Hourihane, Developmental pathways in food allergy: a new theoretical model, *Allergy*, 2009, **64**, 560–568.
40. A. DunnGalvin, *et al.*, Preliminary Development of the Food Allergy Coping and Emotions Questionnaires for Children, Adolescents, and Young People: Qualitative Analysis of Data on IgE-Mediated Food Allergy from Five Countries, *J. Allergy Clin. Immunol. Pract.*, 2018, **6**(2), 506–513.

41. H. Monks, M. H. Gowland, H. MacKenzie, M. Erlewyn-Lajeunesse, R. King, J. S. Lucas and G. Roberts, How do teenagers manage their food allergies?, *Clin. Exp. Allergy*, 2010, **40**(10), 1533–1540.
42. K. J. Allen, P. J. Turner and R. Pawankar, *et al.*, Precautionary labelling of foods for allergen content: are we ready for a global framework?, *World Allergy Organ. J.*, 2014, 7(1), 10.
43. S. S. Sheth, M. Ben-Shoshan, D. W. Harrington, J. Fragapane, L. Soller, L. Joseph St., Y. Pierre, S. La Vieille, S. J. Elliott, S. Waserman, R. Alizadehfar, L. Harada, M. Allen, M. Allen and A. E. Clarke, Effect of precautionary statements on the purchasing practices of Canadians directly and indirectly affected by food allergies, *J. Allergy Clin. Immunol.*, 2012, **129**, 1401–1404.
44. (a) G. Zurzolo, J. J. Koplin, M. L. Mathai, M. K. L. Tang and K. J. Allen, Perceptions of precautionary labelling among parents of children with food allergy and anaphylaxis, *Med. J. Aust.*, 2013a, **198**, 621–623; (b) K. M. Brown, N. E. Fenton, L. D. Lynd, C. A. Marra, J. M. FitzGerald, S. S. Harvard, M. Rosenthal, B. Y. L. Chow, A. E. Clarke and S. J. Elliott, Canadian policy on food allergen labelling: consumers' perspectives regarding unment needs, *Univ. J. Public Health.*, 2015, **3**, 41–48.
45. M. A. Sampson, Risk-taking and coping strategies of adolescents and young adults with food allergy, *J. Allergy Clin. Immunol.*, 2006, **117**(6), 1440–1445.
46. T. Stojanovic, H. McNae, P. Tett, T. W. Potts, J. Reis, H. D. Smith, and I. Dillingham, The "social" aspect of social-ecological systems: a critique of analytical frameworks and findings from a multisite study of coastal sustainability, *Ecol. Soc.*, 2016, **21**(3), 15.
47. L. Polloni and I. Baldi, *et al.*, Multidimensional analysis of food-allergic children and adolescents' self-concept: A comparison with a healthy matched sample, *J. Health Psychol.*, 2015, **20**(6), 850–857.
48. M. C. Young, *et al.*, Management of food allergies in schools: A perspective for allergists, *J. Allergy Clin. Immunol.*, 2009, **124**(2), 175–182.e4.
49. A. DunnGalvin and J. O. B. Hourihane, *Health-related Quality of Life in Food Allergy: Impact, Correlates, and Predictors*, Bundesgesundheitsbl, Springer-Verlag Berlin Heidelberg, 2016.
50. K. Larson, *et al.*, Health Disparities: A Life Course Health Development Perspective and Future Research Directions, in *Handbook of Life Course Health Development*, ed. N. Halfon, C. Forrest, R. Lerner and E. Faustman Springer, Cham, 2018.
51. P. Y. Ong, Consumer attitudes and risks associated with packaged foods having advisory labeling regarding the presence of peanuts, *J. Allergy Clin. Immunol.*, 2007, **120**(1), 171–176.
52. *Consumer Study on Food Allergen Labelling: Follow-on Survey 2008-09*, Evaluation Report Series No. 20, Food Standards Australia New Zealand, 2009.
53. R. E. Poms, C. L. Klein and E. Anklam, Methods for allergen analysis in food: a review, *Food Addit. Contam.*, 2004, **21**, 1–31.

54. B. C. Remington, J. L. Baumert, D. B. Marx and S. L. Taylor, Quantitative risk assessment of foods containing peanut advisory labeling, *Food Chem. Toxicol.*, 2013, **62**, 179–187.
55. K. J. Allen, B. C. Remington, J. L. Baumert, R. W. Crevel, G. F. Houben, S. Brooke-Taylor, A. G. Kruizinga and S. L. Taylor, Allergen reference doses for precautionary labeling (VITAL 2.0): Clinical implications, *J. Allergy Clin. Immunol.*, 2014, **133**, 156–164.
56. Blom, *et al.*, Accidental food allergy reactions: products and undeclared ingredients, *J. Allergy Clin. Immunol.*, 2018, **142**, 865–875.
57. P. Muller, *et al.*, Report 2018 for the European Commission, Directorate-General for Internal Market, Industry, Entrepreneurship and SMEs, ed. K. Hope, CARSA, LE Europe, DIW Econ.
58. M. M. Kelleher, J. O. Hourihane, A. Dunngalvin, C. Cullinane, J. Fitzsimons and A. Sheikh, A 24-h helpline for access to expert management advice for food allergy-related anaphylaxis in children: protocol for a pragmatic randomised controlled trial, *BMJ Open*, 2012, **2**, e001282.
59. A. DunnGalvin, G. Roberts, L. Regent, M. M. Austin, F. Kenna, S. Schnadt, A. Sanchez–Sanz, P. Hernandez, B. Hjorth, M. Fernandez–Rivas, S. Taylor, J. Baumert, A. Sheikh, S. Astley, R. Crevel and E. N. C. Mills, Understanding how consumers with food allergies make decisions based on precautionary labelling, *Clin. Exp. Allergy*, 2019, DOI: 10.1111/cea.13479.
60. C. A. Marra, S. Harvard and M. Grubisic, *et al.*, Consumer preferences for food allergen labeling, *Allergy, Asthma, Clin. Immunol.*, 2017, **13**, 19.

CHAPTER 9

Food Labelling Practices in North America: Canada and the USA

SAMUEL BENREJEB GODEFROY*[a] AND KENNETH TENG[b]

[a] Food Risk Analysis and Regulatory Excellence Platform (PARERA), Department of Food Science and Institute of Nutrition and Functional Foods (INAF), Université Laval, 2425 Rue de l'Agriculture, Québec, QC G1R4V6, Canada; [b] REGSCI PTY LIMITED, 71 Peacock Street, Seaforth NSW 2092, Australia
*Email: parera@fsaa.ulaval.ca

9.1 Introduction

North American jurisdictions were amongst the first regulators to require the use of food labels to communicate key nutrition information relevant to the health of consumers. Nutrition labelling was made mandatory in the USA with the adoption of the Nutrition Labelling Education Act (NLEA) of 1990[1] (amending the Federal Food, Drug and Cosmetic Act), which mandated that most food sold in the USA must bear nutrition labelling and that food labels displaying nutrient content claims and certain health messages comply with specific requirements. Since then, the landscape of food and nutrition labelling has changed significantly, not only in the USA but around the world. Shortly thereafter, Canada followed suit with the development of voluntary nutrition labelling guidelines[2] that evolved into regulations in 2002 and were enacted in full in 2007. Nutrition labelling requirements in the USA and Canada (North America) have been the precursors of a

Food Chemistry, Function and Analysis No. 22
Health Claims and Food Labelling
Edited by Siân Astley

Published by the Royal Society of Chemistry, www.rsc.org

movement adopted by other food regulators globally in accordance with the recommendations of the Codex Alimentarius Commission and in particular of standard CAC/GL 2 – 1985.[3] In examining the North American food and nutrition regulatory labelling provisions, there are two main categories:

1. Mandatory labelling requirements that must appear on labels of most prepackaged foods. These are labelling measures imposed as part of risk management approaches to reduce risks such as those related to non-communicable diseases (NCDs, *e.g.* cardiovascular disease, diabetes, obesity). However, it was in this context that nutrition labelling measures or allergen labelling requirements were also instituted.
2. Discretionary labelling decisions adopted by industry to, for example, highlight benefits of food ingredients and their positive impacts on health. These are framed by other regulatory requirements that specify conditions that must be fulfilled for these statements to be utilised as well as any additional imposed measures, such as the types of statements that can be used, any flexibility in wording and guidance on how conditions that must be satisfied can be met. This is the case for health claims as well as discretionary claims for free-from products (*e.g.* gluten).

This chapter summarises reviews of these categories for labelling provisions, as imposed by North American food regulators, namely Health Canada and the Canadian Food Inspection Agency (CFIA) for Canada as well as the US Food and Drug Administration (US FDA) and the US Department of Agriculture (USDA) for the USA.

9.2 Food Labelling Measures Imposed as a Risk Management Approach

Risk management labelling represents a requirement to be shown on most prepackaged foods. Their intent is to alert consumers of an important fact related to the food composition or conditions of preparation that are not evident from other parts of the label, such as the list of ingredients or method of preparation, which may create a different risk profile for the product. This information is critical to support informed choice as to the suitability of the product in relation to health. For example, both Canadian and American food regulators have imposed the inclusion of a statement "contains phenylalanine" on labels for all products that contain aspartame – a sugar substitute regulated as a food additive in both countries. Although aspartame has to be identified in the ingredients list, there is no indication that this substance represents a source of the amino acid phenylalanine. Phenylalanine is an essential amino acid present in nearly all proteins and protein-containing foods, such as meat, dairy, nuts, legumes and eggs. However, for those with the genetic metabolic disorder phenylketonuria (PKU), this amino acid represents a serious health risk, particularly for children. Individuals with PKU are deficient in the enzyme essential for metabolism of phenylalanine and, when not adequately metabolised, phenylalanine builds up in the blood as phenyl pyruvic acid, which leads to adverse effects spanning from muscle pain and loss of skin pigmentation to brain damage.

With aspartame being added to non-protein-containing foods, there is a need to supplement information on products containing aspartame to help individuals with PKU avoid phenylalanine. Additional information on the food label is, therefore, vital to ensure risk reduction. This illustration exemplifies how policy parameters drive the imposition of food labelling requirements, as a risk management tool in North America, but the health rationale must be well substantiated and the required information should not otherwise be accessible to those consumers needing to manage the established risk. Also, the information is not a warning describing a health repercussion, but rather a statement about composition, on the basis of which decisions can be made. Moreover, in North American jurisdictions where freedom of speech is a well-guarded constitutional principle, space available on a food label may be considered part of the ways in which such freedom is exercised by the food manufacturer. Thus, restricting space with regulatory requirements for mandatory labelling has to meet a high threshold of justification, particular with respect to the health rationale.

Three examples in this chapter are used to illustrate this approach:

- Nutrition labelling
- Mandatory labelling of mechanically tenderised beef
- Allergen labelling

9.2.1 Nutrition Facts Labelling

Nutrition labelling for most prepackaged foods is a mandatory legal requirement in Canada (*via* the Food and Drug Regulations enabled by the Food and Drugs Act) and in the USA (through the Nutrition Labelling and Education Act of 1990). The Canadian Nutrition Facts table or the US Nutrition Facts label organise nutrient content into simple, consistent layouts to help consumers make informed food choices and, possibly, contribute to managing the burden of diet-related chronic diseases of public health significance.

The health rationale that was advanced to justify these mandatory measures was related to reduction of risk factors associated with NCDs and was part of a train of measures being considered internationally to curb increases in the burden of NCDs during the early 2000s. Such efforts culminated with adoption, by the World Health Organization (WHO) in 2004, of a global strategy on diet, physical activity and health, after more than two years of consultations and development.[4] The WHO strategy responds to statistics showing that there are over 1 billion overweight people in the world and at least 300 million clinically obese individuals.[5] The 2002 World Health Report[6] identified that five of the top 10 global disease burden risk factors, namely high blood pressure, high cholesterol, low intakes of fruit and vegetables, high body mass index and physical inactivity, are also major causes of NCDs. Increased exposure to foods generally high in fats, sugar and salt and limited access and affordability of fruits and vegetables are large contributors to these risk factors and represent the impetus behind making improved nutrition information readily available on food labels as a viable risk management option.

North American food regulators used this rationale to mandate that nutrition information be available on food labels, with the focus on identifying nutrients with public health impacts, namely those that needed to be decreased in the diet (*e.g.* fat, sugar and salt) as well as micronutrients that may be associated either with a deficiency of intake in the population or some of its subsets (*e.g.* vitamin D and calcium) or with requirements of management for individuals with specific conditions, where the diet may have an influence (*e.g.* iron and potassium).

Nutrition labelling requirements are considered amongst the most stringent food labelling measures: they are directive both as to what needs to be labelled (supported by a robust health rationale) and as to how it should appear on the label. As a result, North American food regulators tend to be very thorough in justifying judicious selection of the policy option through a mandatory cost–benefit analysis, which is published with the proposed regulatory changes. For example, Health Canada's latest nutrition labelling requirements, published in December 2016, included an economic analysis that estimated changes in labelling requirements would result in benefits reaching 1.926 billion Canadian dollars over 10 years, assuming a 1% reduction (compounded annually) in direct and indirect health costs. These costs are identified as being associated with five chronic diseases (*i.e.* cardiovascular disease, malignant neoplasm, diabetes mellitus, musculoskeletal disease and nutritional deficiency). It is assumed that consumers, when given better information to make food choices, will make healthier decisions, which could in turn reduce negative health outcomes. The costs associated with implementation of the imposed measures over a period of five years were estimated at C$554.7 million, which would make the anticipated net benefit C$1.371 billion over 10 years.[7]

Information required for display on prepackaged food labels for the Nutrition Facts table or NFt (Canada)/Nutrition Facts label (USA) are:

- Serving size (Canada)/servings (USA)
- Calories
- % daily value
- Information on 13 core nutrients:
  - fat (Canada)/total fat (USA)
  - saturated fat
  - trans fat
  - cholesterol
  - sodium
  - carbohydrate (Canada)/total carbohydrate (USA)

- fibre (Canada)/dietary fibre (USA)
- sugars
- protein
- select vitamins
- calcium
- iron
- % daily value footnote (Canada)/footnote (USA)

9.2.1.1 *The Canadian Nutrition Facts Table*

The December 2016 food regulatory amendments included enhancements in the requirements of nutrition labelling, namely:

- Serving size and caloric information were made more accessible with increased font size in addition to the inclusion of a bold line directly beneath.
- Serving size was made more consistent, facilitating comparison between similar foods, and realistic, representing the amount a typical Canadian consumes in one sitting. Imposed reference amounts were published in the regulations, reducing the flexibility that industry had enjoyed before, which was sometimes misused to misrepresent information on food labels.
- Percent (%) daily values were updated based on the latest scientific evidence. In particular, the reference value for sodium was set at 2400 mg/day.
- A new daily value of 100 g was set for total sugars to support calculation of % daily values.
- The list of nutrients was revised to:
  - include potassium, given its importance for maintaining a healthy blood pressure;
  - remove vitamins A and C, as most Canadians are not deficient in these micronutrients;
  - have amounts of potassium, calcium and iron added in mg;
  - require a footnote to be introduced about the way % daily value can be used to interpret nutrition information to aid consumers in understanding nutrient content, *e.g.* 5% or less is a little, 15% or more is a lot.

Industry was given five years from December 2016 to modify labels and adopt the new requirements (see Figure 9.1).

9.2.1.2 *The US Nutrition Facts Label*

The US FDA introduced the new Nutrition Facts label for prepackaged foods on 20 May 2016. On 27 May 2016, the final rules were published in the Federal Register.[9] Figure 9.2 outlines key changes to the Nutrition Facts label.

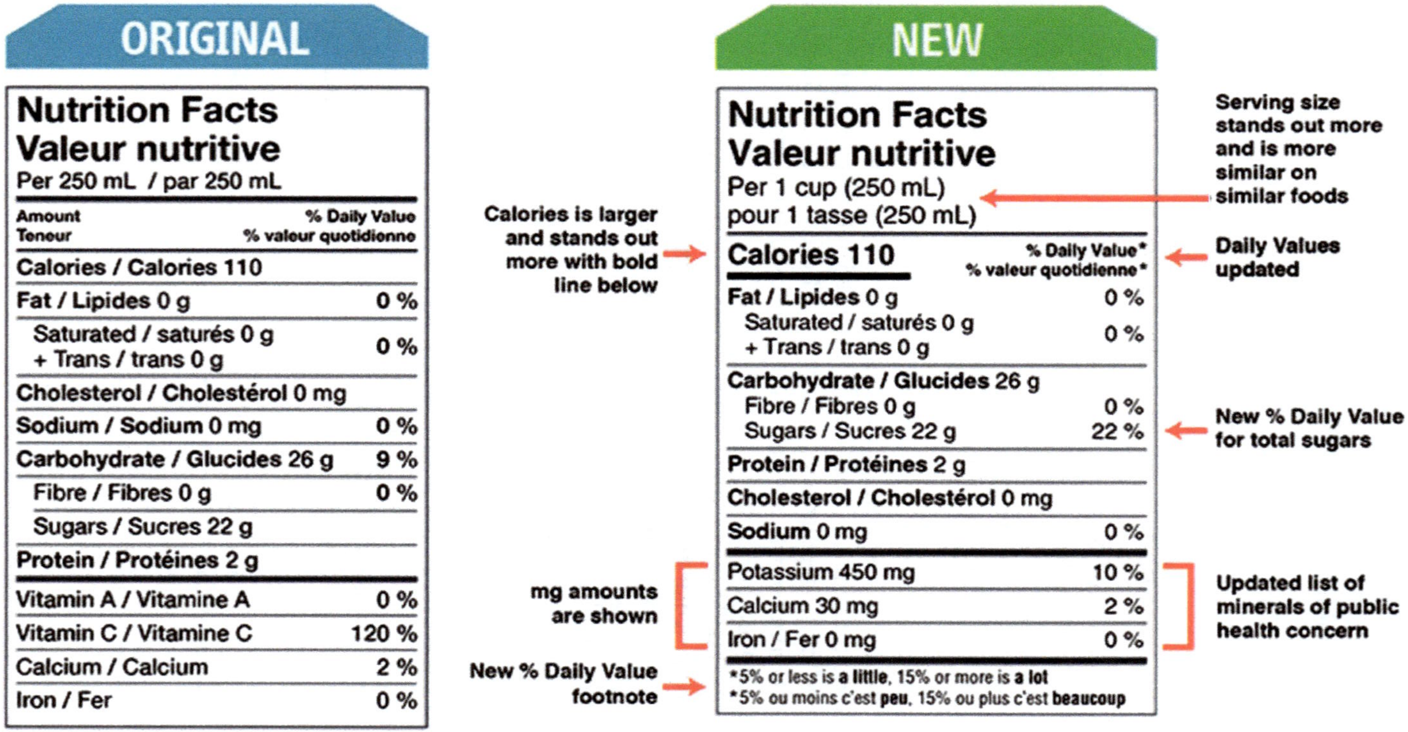

Figure 9.1 Canadian Nutrition Facts table with key changes highlighted.[8] Food Labelling Changes. Health Canada, modified 2017.
Adapted and reproduced with permission from the Minister of Health, 2019.

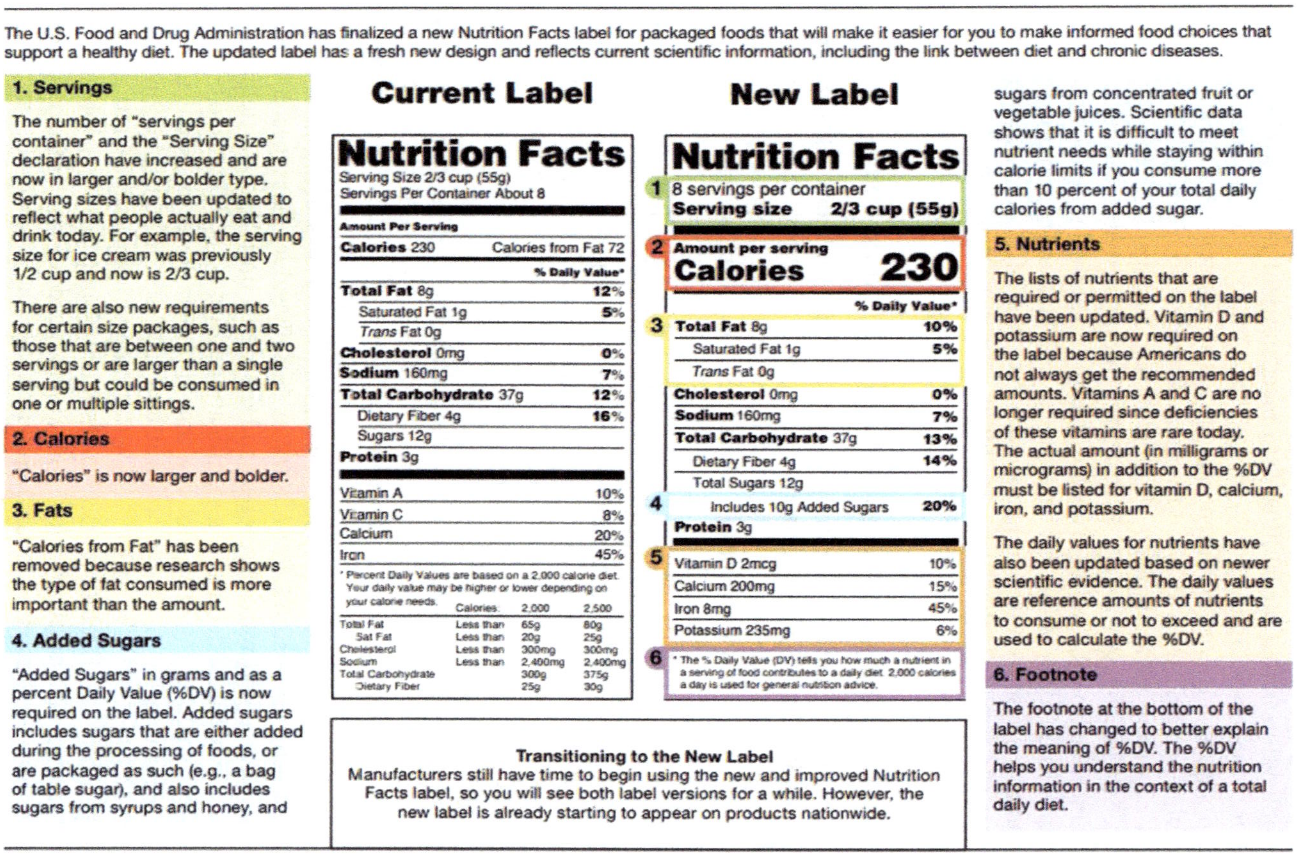

Figure 9.2 US Nutrition Facts label with key changes highlighted. Courtesy of the US Food and Drug Administration.[10]

Some of the key differences in the newly imposed US requirements included the obligation to declare added sugars, as opposed to total sugars, as a % daily value, based on a daily value of 50 g. The rationale advanced for this obligation is that "it is difficult to meet nutrient needs, while staying within the calories limits, if 10% of the total daily calories is coming from added sugars". This requirement was imposed despite the absence of accessible analytical tools that discriminate between added sugars and naturally occurring sugars already in the food (*e.g.* fructose in fruit, lactose in milk). Similar changes were also made for micronutrient declaration, with the removal of the obligation to declare vitamins A and C and imposition of a % daily value declaration for vitamin D, calcium, iron and potassium. A footnote is also required to explain the meaning of % daily value in the context of a total daily diet. Some legibility changes were also made, such as increased font size for the declaration of calories.

9.2.1.3 Front-of-pack Nutrition Labelling Information

9.2.1.3.1 Emergence of Front-of-pack Labelling Practices: Health and Policy Rationale. In view of continued pressure on public health systems related to NCDs, governments and food regulators, in particular, maintained their quest for effective interventions to curb the increasing impacts of such conditions, given the established link between poor diet and increased risk of NCDs. Nutrition labelling information presented on the back of food packages is regarded as requiring a satisfactory level of health literacy with, at minimum, the ability to interpret and utilise qualitative and quantitative data displayed (*e.g.* serving sizes and percent daily values [% DV]). Thus, education initiatives were launched to increase the ability of consumers to use nutrition information featured predominantly on the back of labels.

An example of such a campaign is the collaborative effort carried out by one of Canada's food regulators (Health Canada) in coordination with a coalition of food manufacturers represented by the Food Consumer Products of Canada (FCPC),[11] the Nutrition Facts Education Campaign, which was initiated in 2010 and targeted dissemination of messaging in relation to promoting understanding of % DV or % daily value. Messages deemed easy to understand were developed and displayed on labels. Participating food manufacturers dedicated a significant portion of food labels to convey messages such as a % DV showing that numbers lower or equal to 5% correspond to a low presence of a nutrient (5% is a little) or, in contrast, levels of 15% or above equate to a significant or a high presence in the food composition (15% is a lot). Key messages were accompanied by explanatory information. For nutrients that consumers need to consider reducing their intake of (fat, sugars and sodium), 5% or less can be targeted whilst products indicating 15% or higher can be avoided. Similarly, for those nutrients that may be desirable to address a deficiency, 15% or more can be targeted. A web page was devoted to the campaign (https://www.canada.ca/en/health-canada/services/understanding-food-labels/percent-daily-value.html) but the campaign relied not only on the food package information, provided by participating industry partners, but also on a media plan involving broadcast messages and consumer resource information (*e.g.* factsheets, FAQs) made available as downloadable material and in-store, and support through better product placement, including additional display elements in selected stores.[12]

Evaluation of impact in reference to the educational programmes associated with deployment of the Nutrition Fact Panels (NFP) was carried out including a systematic review by Moore *et al.*[13] Most studies examined were conducted in the USA and reported a statistically significant improvement in one or more outcomes of participant understanding or use of nutrition labels. The review confirmed the importance of nutrition label literacy together with education on general healthy eating recommendations. These instructional efforts supported the effectiveness of nutrition labelling interventions and resulted in impacts on label use and subsequent consumer behaviours. A Canadian study chosen as part of this review used in-store education with the participation of a registered dietician[14] and reported increased ability to use nutrition labels and better understanding in relation to "serving size and percent daily value". However, a Canadian study[15] focusing on evaluation of impacts of the Canadian Nutrition Facts Education Campaign amongst young Canadians concluded that a mass media campaign would not be sufficient to enhance population level understanding of the % DV.

The need to make nutrition information more readily available, accessible and usable has led to increased reliance on simplified and interpretive nutrition labels, presented on the front of packaging, so-called front-of-pack (FOP) labelling systems. Canada's food regulators have made a regulatory proposal requiring FOP labelling on a selection of targeted food products. Acton *et al.*[16] suggested that Canadian consumers would support such measures unanimously and that the expected impacts would result in improved communication about the "healthiness" of beverage products with a less obvious nutrition profile, such as chocolate milk.

9.2.1.3.2 Canada's Proposed Mandatory Front-of-pack Labelling Requirements. Health Canada introduced new *Regulations amending the Food and Drug Regulations (Nutrition Symbols, Other Labelling Provisions, Partially Hydrogenated Oils and Vitamin D)* with the intent to impose mandatory nutrition FOP labelling. The *Proposed Regulations* were published in the Canada Gazette Part I: Vol. 152, No. 6 on 10 February 2018. This mandatory nutrition FOP labelling targets foods high in saturated fat, sugars and sodium with the goal of reducing the health risks, specifically obesity, heart disease and high blood pressure, in Canada. The proposed nutrition FOP labelling amendments included:

- FOP nutrition symbol: four design options have been developed (see ref. 17 for details).

- Thresholds: requirement to include FOP nutrition symbol when specific thresholds are met or surpassed. For prepackaged foods and foods destined for children only (aged 1 to 4 years), the proposed limit is 15% DV for saturated fat, sugars and sodium.
- Exemptions: foods that are not required to use a FOP nutrition symbol include those already exempt from bearing a NFt, foods with protective health effects (*e.g.* fresh, frozen or canned fruits and vegetables), and those where including FOP details would be redundant (*e.g.* table salt and sweetening agents, such as sugar and honey).
- Format specifications are meant to ensure FOP nutrition symbol is obvious, legible and user-friendly.

The imposition of such requirements comes at a time where international discussions are under way, as part of the Codex Alimentarius Commission and its subsidiary committees, as to relevance and requirements associated with reliance on these simplified nutrition labelling provisions. Similarly, and while results from some studies seem to suggest consumer preference for these simplified tools, demonstration of their effectiveness, as suggested by the Canadian regulations, has not been made with the application of voluntary mechanisms and a thorough evaluation of expected impacts. Reliance on the regulatory instrument, in this instance by Canadian food regulators, represents a departure in policy direction thus far, where strength of the health rationale, progressive implementation – from voluntary to compulsory measures – and demonstration of effectiveness are generally pivotal to advance a comprehensive mandatory provision.

Moreover, these regulatory measures do not factor in ongoing changes in trends in consumer purchasing behaviours, including increased use of home-delivered foods and online grocery shopping, where FOP information is not effective. Similarly, the role of technology and use of smart phones, as a key source of information and decision-making for day-to-day activities, have not been considered in relation to food labels as the primary source of information in making food purchasing decisions. Numerous applications are being designed to combine information extracted from bar codes associated with prepackaged products and to deliver personalised recommendations as to the suitability of a food product for individuals with specific nutritional requirements. These measures may be far more effective and less costly than implementation of mandatory FOP requirements.

9.2.2 Mandatory Labelling of Mechanically Tenderised Meat Products

Mechanically tenderised meat products are a good illustration of the application of mandatory labelling measures resulting from a public health concern. The issue being managed is microbiological risk, which has the potential to alter the risk profile of a product, hence the provision of additional information for consumers to take precautions in handling and cooking.

9.2.2.1 *Canadian Requirements*

On 21 May 2014, Health Canada imposed mandatory labelling of mechanically tenderised beef (MTB),[18] including safe cooking instructions. These requirements emerged as a result of health concerns stemming from the fact that mechanical tenderisation is a common practice that may increase the risk of harmful bacteria being transferred from the surface to the centre of the meat.[19] As a result, cooking advice, regarding temperatures to inactivate these bacteria, was adapted to accommodate this new risk profile.[20] In June 2014, Health Canada also issued *Guidance on Mandatory Labelling for Mechanically Tenderized Beef*[21] to provide industry with information on labelling requirements of MTB sold in Canada.

Increased risk to consumers from MTB was established through a risk assessment performed by scientists from Health Canada in 2013.[22] It was also identified that consumers were not able to differentiate between uncooked regular meat cuts and others subjected to tenderisation and labelling would be a suitable tool to address this gap. The health rationale was further bolstered by eight cases of food-borne illness caused by *Escherichia coli* (*E. coli*) 0157 where five cases were identified as likely to be associated with the consumption of beef that had been mechanically tenderised by the retailer.

Canadian regulations require that consumers are advised about safe cooking conditions in a consistent fashion and in a manner that is legible and as prominent as other informational text on the principal display panel. Overall, the regulation stipulates that:

1. The wording "mechanically tenderised" appears as part of the common name of the product or elsewhere on the principal display panel.
2. This wording be legible and prominent in the same fashion as the common name.
3. Cooking interactions appear on the principal display panel: "Cook to a minimum internal temperature of 63 °C or 145 °F".
4. The instruction to turn steak over at least twice during cooking appear with the cooking directions if the product is a steak.

These requirements are adapted for non-prepackaged products, such that the wording "mechanically tenderised" is present on any label associated with the product (*e.g.* in-store signs placed adjacent to the product in the display case). Cooking instructions are added to the principal display panel on the label used once the product is packaged for sale.

9.2.2.2 US Requirements

The USDA proposed labelling requirements for beef products that are mechanically tenderised with a needle or blade invoking the same rationale:

- Potential of pathogen transfer from the exterior of the product into the interior, increasing its health risk profile.
- Inability to distinguish mechanically tenderised products from non-tenderised equivalents.

The risk considerations were motivated by outbreak data[23–25] indicating that consumers in food service establishments would not cook mechanically tenderised products to a temperature and/or for sufficient time to destroy effectively harmful bacteria that might have transferred to the interior of the product. USDA regulatory requirements were published in the Federal Register in 2015[26] and the regulations, which became effective on 17 May 2016, call for the use of the descriptive designation "mechanically tenderised" or an equivalent term, such as "blade" or "needle" tenderised on the labels of raw or partially cooked products that have been mechanically tenderised.

The information is not required if the product is intended to receive another treatment, such as cooking or full lethality treatment, at an official establishment designated by the regulator.

The US requirements also mandate addition of what is termed "validated" cooking instructions, with flexibility as to their determination. Primarily, guidance is provided to achieve such validation, along with some examples. In providing further justification as to the judicious choice of this risk management approach, the USDA cited studies that showcased a change in consumer behaviour as a result of availability and accessibility of safe handling instructions on food labels.[27] Studies and data showcasing possible public health repercussions of consuming products that have been mechanically tenderised, and not been cooked properly at food service establishments, were used to strengthen the public health rationale.

9.2.3 Mandatory Food Allergen Labelling Measures

Food allergy prevalence in Canada is estimated to be 7.14% in children and 6.56% in adults,[28] although the authors indicated this prevalence relied on self-reported conditions, which is likely to lead to possible overestimation. Nevertheless, US statistics, reported through similar studies, found that food allergy affected 6–8% of infants and young children and 3.5–4% of adults.[29,30]

Food allergy incidents can lead to mild symptoms, such as itching or urticaria, and to anaphylaxis and death. The health rationale for using food labelling as an instrument to prevent food allergy incidents is well established; at present, food allergies cannot be cured and the only management approach is avoidance of the ingredient to which the allergic individual is likely to react. Food labels can be the only link between allergic consumers or their carers and food product manufacturers. Thus, labels can offer valuable information on the composition of prepackaged products, including ingredients that individuals might wish to avoid. Most prepackaged foods in North America are mandated to bear an ingredient list and provisions for food allergy, enacted through regulations in Canada[31] and legislation in the USA,[32] focused on enhancing existing ingredient labelling requirements.

9.2.3.1 Labelling of Allergens as Part of the List of Ingredients

Canada and the USA have adopted a list of priority food allergens, based on the list established by Codex,[33] with minor revisions for the USA and some additions for Canada. Priority food allergens in Canada include cereals containing gluten (wheat, rye, barley, oats, spelt or their hybridised strains and products of these), crustaceans and products of these, egg and egg products, fish and fish products, peanuts, soybeans, milk and milk products, tree nuts and nut products, mustard, and sulphites (10 $mg\,kg^{-1}$ or higher). The Canadian definition of food allergens includes reference to the protein fraction, *i.e.* "any protein from any of the following foods or any modified protein, including any protein fraction that is derived" from the identified list of priority foods. This definition was part of regulatory amendments to the Canadian Food and Drug Regulations that became effective in 2012 and focused on enhancing requirements for labelling when priority ingredients are added deliberately.

In the USA, priority food allergens consist of milk, egg, fish (*e.g.* bass, flounder and cod), crustacean shellfish (*e.g.* crab, lobster and shrimp), tree nuts (*e.g.* almonds, walnuts and pecans), wheat, peanuts and soybeans. These food allergens, or any protein derived from them, are designated as "major food allergens" by the US Food Allergen Labelling and Consumer Protection Act of 2004 (FALCPA), which amended the Federal Food, Drug, and Cosmetic Act, and became effective on 1 January 2006. The law applies to all food regulated by the US FDA. Similar provisions have also been developed for food products regulated by the USDA (*i.e.* meat, poultry and certain egg products). In both instances, the intent of the enhanced labelling requirements is to ensure that allergens are not overlooked by allergic consumers. The requirements call for identification of ingredients using systematic wording that is easy for consumers to understand, with the option to incorporate a "contains statement" that lists all priority allergens that are present on a separate line.

The regulatory requirements also called for the removal of any labelling exemptions that may have applied to food allergens if they formed part of a multi-ingredient product, such as spices and seasonings, which are exempt from sub-ingredient declaration. The intent was to ensure that consumers have more certainty about food composition, thereby effectively eliminating any concerns related to missing ingredients that might prove detrimental to health. In this case, the policy rationale was well corroborated, *i.e.* health benefits are clearly generated and/or negative health impacts prevented through the implementation of mandatory allergy labelling requirements. Moreover, these labelling measures were integrated with the overall ingredient labelling requirements and do not represent a departure from them.

9.2.3.2 Allergen Precautionary Labelling

In both countries, allergen labelling provisions did not cover compulsory declaration of allergens present in food as a result of non-deliberate, yet unavoidable, cross-contamination or commingling of products, *e.g.* commingling of cereals and soy products. These situations continue to be managed with the use of voluntary precautionary statements.[34] Canadian food regulators were amongst the first to recommend the use of precautionary statements, as a means to alert consumers to the possible presence of undeclared allergens, as a result of cross-contamination. General conditions for the use of such statements have been developed widely since the 1990s. Health Canada defined a food allergen precautionary statement as a "declaration on the label of a prepackaged food of the possible inadvertent presence of an allergen in the food". When used, precautionary statements are meant to deter allergic consumers from consumption of the product, as the presence of harmful levels of allergenic ingredient(s) cannot be ruled out. Health Canada's policy has been non-prescriptive with respect to the wording of precautionary statements, requiring only that "such statements be truthful, clear and non-ambiguous, and that they not be a substitute for Good Manufacturing Practices". The proliferation of these statements on food labels, in a manner that does not suggest a risk-based approach, using a "may contain" statement that encompasses most, if not all, of the list of priority allergens, has contributed to trivialisation of the message that was to be conveyed by these statements and critically weakened their risk management potential. Health Canada and the CFIA have attempted to strengthen guidance related to the use of such statements, with a recommendation to use one phrase "may contain [X]", rather than multiple statements describing manufacturing conditions, such as "produced in a facility that uses [X]", "produced on the same line where [X] is used". Additionally, the voluntary nature of these requirements has not contributed to any improvement in their effectiveness nor increased trust in their value as deterrents to consumption. The situation is the same for regulatory provisions applied to allergen precautionary statements in the USA, where only voluntary guidance was found to support oversight of such labelling practices.

Regulators in both countries supported recent proposals made by the Codex Committee on Food Labelling (CCFL) to develop international guidance on precautionary labelling measures.[35] It is expected that, with the progression of this work and the possible development of threshold-based guidance on conditions of use for precautionary statements, changes will be triggered in associated Canadian and US regulatory requirements, as well as elsewhere globally, with a probable shift towards mandatory requirements over the next three to five years.

9.3 Discretionary Food Labelling Measures Framed with Regulatory Requirements

9.3.1 Health Claims

Health claims are being used increasingly on food labels to alert consumers to the potential health benefits, as a result of the presence or the addition of food ingredients considered functional, *i.e.* having an additionally beneficial health-related effect beyond nutrition. Consumers continue to have a keen interest in learning more about their food choices, including the scientific evidence linking ingredients to health-enhancing properties, and are taking the initiative in selecting appropriate food products for their health status.

Health claims are explicitly defined in US legislation, "Health claim means any claim made on the label or in labeling of a food, including a dietary supplement, that expressly or by implication, including "third party" references, written statements (*e.g.* a brand name including a term such as "heart"), symbols (*e.g.* a heart symbol), or vignettes, characterises the relationship of any substance to a disease or health-related condition."[36] However, in Canadian regulations, this term has not been defined formally but rather the Codex Alimentarius definition has been adopted by policy, *i.e.* "a health claim is defined as any representation in labelling and advertising, that states, suggests or implies that a relationship exists between the consumption of a food or a component of that food and health".[37] According to these definitions, both US and Canadian regulators consider that claims signify more than mere statements on food labels, but also encompass implied expressions through slogans, graphics, logos, symbols or other means, such as a name or even a trademark.[38]

It is possible to consider health claims according to a risk paradigm, as depicted in Figure 9.3.

The risk factor is considered an important component of the claim; a function claim, describing an association between an ingredient and a health effect, is considered to be of lower impact than a therapeutic claim, which associates the ingredient with treatment of a condition or ailment,

Figure 9.3 Health claims risk paradigm.

if the claim is not fulfilled. Similarly, within each claim category, the risk may be different, due to variations in negative consequences with respect to the claim not being achieved, *e.g.* a claim about teeth whitening is of lesser impact than treatment of a condition such as heart disease.

Canada and the USA have established regulatory frameworks to manage health claims. Food law in both countries prohibits misrepresentation of food products, which imply that health claims must be truthful and not misleading. Also, any health claim associated with serious health conditions or disease – such as heart disease, dementia and obesity – trigger the requirement to comply with regulatory provisions for therapeutic products, which are subject to compulsory premarket clearance and registration. This means that a "food" could be classified as a "drug" should its label bear a health claim mentioning some (for Canada) or all (for the USA) diseases. This obligation can, however, be lifted through a regulatory decision by Health Canada or the US FDA under certain conditions. Table 9.1 summarises the Canadian and US regulatory requirements for different classes of health claims.

Another notable difference is the concept of a "qualified health claim", whereby the regulator permits use of a claim, but mandates its qualification, as regards the scientific rationale, *e.g.* claims such as "selenium may reduce the risk of certain cancers". For example, the US regulator allows the use of qualified statements, such as "some scientific evidence suggests that consumption of selenium may reduce the risk of certain forms of cancer; however, the FDA has determined that this evidence is limited and not conclusive". In contrast, Health Canada has developed guidance outlining the standards of evidence that must be met for disease risk reduction and therapeutic claims,[39] including instances when functional claims related to Schedule A diseases, thus necessitating mandatory clearance or not. As such,

Table 9.1 Summary of the regulatory requirements for health claims in Canada and the USA.

Type of claim	Canada	USA
Function claim	• Must be truthful and not misleading • No premarket clearance required	• Must be truthful and not misleading • No premarket clearance required
Disease risk reduction claim	• Must be truthful and not misleading • Premarket clearance for claims related to Schedule A diseases noted in the Food and Drugs Act	• Must be truthful and not misleading • Premarket authorisation required
Therapeutic claims	• Must be truthful and not misleading • Premarket clearance for claims related to Schedule A diseases noted in the Food and Drugs Act	• Not permitted

and in practice, the regulatory environment surrounding health claims is seen by many stakeholders to be significantly more stringent in Canada than in the USA.

With continued investment in studies examining health effects of food ingredients and components, and shifts in consumer interests, habits and behaviours towards food labelling or surrogate information, such as that obtained through smartphone applications, the regulatory environment as related to health claims must continue to evolve, not only in Canada and the USA, but also in other jurisdictions (*e.g.* EU, Australia and Japan) to achieve greater alignment of decisions and resource sharing in assessing the validity of claims sought by industry.

9.3.2 Gluten-free/Allergen-free Claims

Since the introduction of gluten-free food to North American markets over two decades ago their availability on grocery store shelves has increased steadily. Regulators in the USA and Canada were amongst the first to adopt a set of regulatory requirements that framed use of the statement "gluten-free" for foods identified as such by manufacturers who made additional efforts to either remove sources of gluten or replace them with non-gluten-containing ingredients.

Canada's regulations for gluten-free foods was included as part of the regulatory requirements for products with special dietary purposes (division 24 of the Canadian Food and Drug Regulations). This division was amended in 2012 to restrict usage of the statement "gluten-free" for food where the protein portion of select cereals, designated as "gluten" by the regulation is absent, namely "barley, oats, rye, triticale or wheat, kamut or spelt" or any modified protein derived from these grains or the "grain of a hybridised strain created from at least one" of these same cereals.[40] Prior to 2012, the exclusion covered the cereal in its entirety and not exclusively the protein fraction. The amendment was a result of adapting the definition of gluten-free food to continuously evolving knowledge regarding ingredients responsible for the symptoms experienced by those with coeliac disease in response to ingestion of gluten-containing food.

People with coeliac disease react negatively to the consumption of specific components of certain cereal grains. Prolamins, the alcohol-soluble fractions of wheat (gliadins), rye (secalins) and barley (hordeins) are of greatest concern to coeliac individuals.[40] Glutelins, the water-soluble fraction, although abundant in the protein portion of these cereals, appear to be less harmful.[41] Avenins (prolamins from oats), structurally different from the other prolamins, are less prevalent in the overall protein composition of oats (10–15%) and tend to be safe for the majority of those with coeliac disease. However, as there is no cure for coeliac disease, the only measure of protection for these individuals is to adopt a lifelong diet excluding all gluten sources. For such consumers, "gluten-free" statements on food labels are a tool that help identify foods suitable or unsuitable for them.

In Canada and the USA, foods that have been altered to achieve the status of "gluten-free" may indicate so on their label with the condition that no gluten marker above 20 ppm is detected using an analytical method known as the "R5" method. This is in reference to an immunoassay-based technique where the antibody was generated against the pentapeptide region common to all gluten prolamins (except avenins). The 20 ppm threshold was established on the basis of scientific assessment showing that, even in a worst-case scenario where all gluten-free foods were set to a 20 ppm concentration for cross-contamination, exposure to gluten would not represent a health concern for coeliac individuals.[42,43]

Food regulators in both countries have also adopted the policy that "gluten-free" claims cannot be added to foods that are inherently gluten-free, without indicating so. For example, a bottled water product cannot bear a "gluten-free" claim without the clarification that all bottled water products are naturally/inherently gluten-free. Additionally, regulators have endorsed pure and uncontaminated oats (free from other sources of protein) as a gluten-free ingredient, so long as the level of gluten-containing cereals does not exceed 20 ppm, based on detection of marker proteins. In contrast, highly processed beer, which has undergone enzymatic processes aimed at reducing the availability of protein fragments known to elicit immune-mediated reactions associated with coeliac disease, has received mixed evaluations from regulators. Health Canada has acknowledged that a gap in evidence exists as to the safety of these products, given that some protein fragments attributed to gluten can still be detected with methods other than R5 assays, *e.g.* mass spectrometry.

"Allergen-free" statements are not governed with the same regulatory clarity. They are subjected to general provisions of food laws in Canada and the USA, meaning that food labels must be truthful and not misleading. As such, reliance on "allergen-free" statements (*e.g.* "peanut-free") must be supported by specific measures, employed by food processors, to ensure the allergen is not included in the formulation of, nor come in contact with, the food product. Such efforts must be sustained over time. North American food regulators can, however, target allergen-free products, more than others, through specific compliance and enforcement campaigns, due to their higher risk profile. Allergic consumers and their care-givers tend to characterise "allergen-free" products as safe; failure in the fulfilment of such claims has more significant impact, in view of the fact that food allergy can lead to anaphylaxis and death. Recalls of possible non-compliant products are frequent in Canada and the USA.

Based on the perpetual progress of food processing techniques and the additional interest to diversify gluten-free and allergen-free food offerings, it is expected that this area will remain highly active with continued innovation to gluten-free and allergen-free labels and their conditions of use, with a strong possibility for enhanced regulatory oversight for allergen-free products.

9.4 Conclusion

The food regulatory environment surrounding labelling requirements in Canada and the USA has evolved over time, marked by instances where it has led the way internationally, initiating new work by the Codex Alimentarius Commission, and addressing emerging needs, such as continued and progressive consumer interest to know more about the food products they consume and possible health effects or standards of evidence that must be fulfilled to substantiate specific decisions requiring mandatory labelling (*e.g.* FOP labelling). Considering the abundance of food products traded between both countries, there is an urgent need for even greater alignment of food labelling policies. Similarly, and while a number of policy levers for certain food labelling requirements are documented with specific regulatory decisions, there is a benefit for principles guiding and framing such policies to be articulated clearly, debated and, possibly, adopted with the aim of ensuring greater consistency and predictability in the food labelling regulatory landscape governing Canada and the USA.

Competing Interests Declaration

Prof. Godefroy acts a food regulatory advisor to a number of international initiatives on food safety (regulatory) capacity building, as well as for some food regulators and food industry manufacturers.

Contributions

All authors listed in the manuscript contributed to the conception, acquisition, analysis and interpretation of data, design and critical revision of the manuscript, including approval of the final submitted version.

References

1. F. R. Shank, The Nutrition Labelling and Education Act of 1990, *Food Drug Law J.*, 1992, **47**, 247–252.
2. Government of Canada. Food and Drug Regulations Part B, Foods Section B.01.401 – B.01.467, Nutrition Labelling. Available from: https://laws-lois.justice.gc.ca/eng/regulations/C.R.C.,_c._870/page-15.html#h-19.
3. Codex Alimentarius, Guidelines on Nutrition Labelling, CAC/GL 2-1985, Last revision 2017. Available from: http://www.fao.org/fao-who-codexalimentarius/sh-proxy/en/ ? lnk=1&url=https%253A%252F%252Fworkspace.fao.org%252Fsites%252Fcodex%252FStandards%252FCAC%2BGL%2B2-1985%252FCXG_002e.pdf.
4. A. Waxman, WHO's global strategy on diet, physical activity and health, *Scand. J. Nutr.*, 2004, **48**(2), 58–60.

5. WHO, Obesity: Preventing and managing the global epidemic, Report of a WHO consultation, WHO Technical Report Series, No. 894, 2000.
6. WHO, Reducing Risks, Promoting Healthy Life, Geneva, World Health Organization, World Health Report, 2002.
7. Cost Benefit analysis, Canada Gazette Part II. Available from http://gazette.gc.ca/rp-pr/p2/2016/2016-12-14/html/sor-dors305-eng.html.
8. Government of Canada, Health Canada, Figure 1 Nutrition facts table changes. Available from: https://www.canada.ca/en/health-canada/services/food-labelling-changes.html#a4.
9. Federal Register, Volume 81, Issue 103, 81 FR 33742 – Food Labelling: Revision of the Nutrition and Supplement Facts Labels (May 27, 2016).
10. U.S. Food and Drug Administration, The New and Improved Nutrition Facts Label – Key Changes, 2018. Available from: https://www.fda.gov/downloads/food/labellingnutrition/ucm511646.pdf.
11. Government of Canada, Health Canada. Available from: https://www.canada.ca/en/health-canada/services/food-nutrition/healthy-eating/sodium/educational-tools-key-messages.html.
12. Government of Canada, Public Works and Government Services Canada. Available from: https://buyandsell.gc.ca/cds/public/2013/06/03/67fe8979dc1010960e7e8e98ba2f4212/appendix_b_-_nfec_evaluation_framework.pdf.
13. S. G. Moore, J. K. Donnelly, S. Jones and J. E. Cade, Effect of Educational Interventions on Understanding and Use of Nutrition Labels: A Systematic Review, *Nutrients*, 2018, **10**, 1432.
14. S. Dukeshire, E. Nicks and J. Ferguson, Pilot Evaluation of an In-Store Nutrition Label Education Program, *Can. J. Diet. Pract. Res.*, 2014, **75**, 206–209.
15. B. Cormier, L. Vanderlee and D. Hammond, Use of Nutrition Information and Understanding of "Percent Daily Value" on Nutrition Facts Tables: Evaluating the Impact of a National Public Education Campaign among Youth and Young Adults in Canada. Available from: https://doi.org/10.3148/cjdpr-2019-010.
16. R. B. Acton, L. Vanderlee and D. Hammond, Influence of front-of-package nutrition labels on beverage healthiness perceptions: Results from a randomized experiment, *Prev. Med.*, 2018, **115**, 83–89.
17. Government of Canada, Health Canada, Summary of proposed amendments published in Canada Gazette, Part I: nutrition symbols, other labelling provisions, partially hydrogenated oils and vitamin D. Available from: https://www.canada.ca/en/health-canada/programs/consultation-front-of-package-nutrition-labelling-cgi/summary-of-proposed-amendments.html.
18. Government of Canada, Food and Drug Regulations. Part B Foods. Division 14 Meat, Its Preparations and Products. Meat, Meat By-Products. B.14.022. Available from: https://laws-lois.justice.gc.ca/eng/regulations/c.r.c.,_c._870/page-68.html.

19. J. B. Luchansky, R. K. Phebus, H. Thippareddi and J. E. Call, Translocation of surface-inoculated *Escherichia coli* O157:H7 into beef subprimals following blade tenderization, *J. Food Prot.*, Nov 2008, **71**(11), 2190–2197.
20. B. A. Smith, A. Fazil and A. M. Lammerding, A risk assessment model for *Escherichia coli* 0157:H7 in ground beef and beef cuts in Canada: Evaluating the effects of interventions, *Food Control*, 2013, **29**, 364–381.
21. Government of Canada, Health Canada, Health Products and Food Branch, Food Directorate. Guidance on Mandatory Labelling for Mechanically Tenderized Beef, 2014.
22. A. Calford, M. C. Lavoie, B. Smith, E. Buenaventura, H. Couture, A. Fazil and J. M. Farber, Findings of the health risk assessment of *Escherichia coli* 0157 in mechanically tenderized beef products in Canada, *Int. Food Risk Anal. J.*, 2013, **3**, 3.
23. L. E. Swanson, J. M. Scheftel, D. J. Boxrud, K. J. Vought, R. N. Danila, K. M. Elfering and K. E. Smith, Outbreak of *Escherichia coli* O157:H7 infections associated with nonintact blade-tenderized frozen steaks sold by door-to-door vendors, *J. Food Prot.*, 2005, **68**, 1198–1202.
24. N. Haubert, A. Cronquist, S. Parachini, J. Lawrence, A. Woo-Ming, T. Volkman, S. Moyer and A. Watkins, Outbreak of *Escherichia coli* O157:H7 Associated with Consuming Needle Tenderized Undercooked Steak from a Restaurant Chain. Presented at the International Conference on Emerging and Infectious Diseases, Atlanta, GA (19–22 March 2006).
25. W. Culpepper, T. Ihry, C. Medus, A. Ingram, D. Von Stein, S. Stroika, E. Hyytia-Trees, S. Seys and M. J. Sotir, Multi-state outbreak of *Escherichia coli* O157:H7, Infections associated with consumption of mechanically-tenderized steaks in restaurants- US 2009. Presented at the International Association for Food Protection, 1–4 August 2010, Anaheim, CA.
26. Federal Register, Volume 30, Number 95, 28153-28172. PR. Doc No2015-11916. Rules and Regulations(18 May 2015).
27. Yang, *et al.*, Evaluation of Safe Food-Handling Instructions on Raw Meat and Poultry Products, *J. Food Prot.*, 1999, **63**, 1321–1325.
28. L. Soller, M. Ben-Shoshan, D. W. Harrington, J. Fragapane, L. Joseph, St., Y. Pierre, S. B. Godefroy, S. La Vieille, S. J. Elliott and A. E. Clarke, Overall prevalence of self-reported food allergy in Canada, *J. Allergy Clin. Immunol.*, 2012, **130**(4), 986–988.
29. J. Grundy, S. Matthews, B. Bateman, T. Dean and S. Arshad, Rising prevalence of allergy to peanut in children: data from 2 sequential cohorts, *J. Allergy Clin. Immunol.*, 2002, **110**(5), 784–789.
30. S. H. Sicherer, Muñoz-Furlong and H. Sampson, Prevalence of seafood allergy in the US determined by random telephone survey, *J. Allergy Clin. Immunol.*, 2004, **114**(1), 159–165.
31. Government of Canada, Regulations Amending the Food and Drug Regulations (Nutrition Labelling, Other Labelling Provisions and Food Colours), Food and Drugs Act, Vol. 150, No. 25 – December 14, 2016.

SOR/2016-305. Available from: http://gazette.gc.ca/rp-pr/p2/2016/2016-12-14/html/sor-dors305-eng.html.
32. US FDA, Food Allergen Labeling and Consumer Protection Act of 2004 (FALCPA). Available from: https://www.fda.gov/food/food-allergens-and-gluten-free-guidance-documents-and-regulatory-information/food-allergen-labeling-and-consumer-protection-act-2004-falcpa.
33. Codex Alimentarius, General Standard for the Labelling of Prepackaged Foods, CXS 1-1985. Last revision 2018. Available from: http://www.fao.org/fao-who-codexalimentarius/sh-proxy/en/ ? lnk=1&url=https%253A%252F%252Fworkspace.fao.org%252Fsites%252Fcodex%252FStandards%252FCODEX%2BSTAN%2B1-1985%252FCXS_001e.pdf.
34. Government of Canada, Health Canada, Health Products and Food Branch, Food Directorate, The Use of Food Allergen Precautionary Statements on Prepackaged Foods, 2012. Available from: https://www.canada.ca/content/dam/hc-sc/migration/hc-sc/fn-an/alt_formats/pdf/label-etiquet/allergen/precaution_label-etiquette-eng.pdf.
35. Codex Alimentarius, Report of the Forty-Fourth Session of the Codex Committee on Food Labelling, REP18/FL (2017). Available from: http://www.fao.org/fao-who-codexalimentarius/sh-proxy/en/?lnk=1&url=https%253A%252F%252Fworkspace.fao.org%252Fsites%252Fcodex%252FMeetings%252FCX-714-44%252FREPORT%252FREP18_FLe.pdf.
36. Code of Federal Regulations, Title 21 Food and Drugs, Volume 2, 2008. Part 101 Food Labeling, Title 21 – Food and Drugs. Chapter 1 – Food and Drug Administration, Department of Health and Human Services, Subchapter B – Food for Human Consumption, Subpart A – General Provisions, Section 101.14 Health claims: general requirements. Available from: https://www.govinfo.gov/content/pkg/CFR-2008-title21-vol2/xml/CFR-2008-title21-vol2-part101.xml#seqnum101.14.
37. Codex Alimentarius, Guidelines for Use of Nutrition and Health Claims, CAC/GL 23-1997. Revised 2004. Available from: http://www.fao.org/ag/humannutrition/32444-09f5545b8abe9a0c3baf01a4502ac36e4.pdf.
38. M. R. L'Abbé, L. Dumais, E. Chao and B. Junkins, Health Claims on Foods in Canada, *J. Nutr.*, 2008, **138**, 1221S–1227S.
39. Government of Canada, Health Canada, Guidance Documents for Preparing Health Claim Submissions. Available from: https://www.canada.ca/en/health-canada/services/food-nutrition/food-labelling/health-claims/guidance-documents-preparing-submission-food.html.
40. Government of Canada, Food and Drug Regulations, Part B Foods, Division 1 General, B.01.010.1. Available from: https://laws-lois.justice.gc.ca/eng/regulations/c.r.c.,_c._870/page-8.html.
41. O. Pulido, M. Zarkadas, S. Dubois, K. Macisaac, I. Cantin and S. La Vieille, *et al.*, Clinical features and symptom recovery on a gluten-free diet in Canadian adults with celiac disease, *Can. J. Gastroenterol.*, 2013, **27**(8), 449–453.

42. A. K. Akobeng and A. G. Thomas, Systematic review: Tolerable amount of gluten for people with coeliac disease, *Aliment. Pharmacol. Ther.*, 2008, **27**, 1044–1052.
43. Government of Canada, Health Canada, Health Canada's Position on Gluten-Free Claims. Available from: https://www.canada.ca/en/health-canada/services/food-nutrition/food-safety/food-allergies-intolerances/celiac-disease/health-canada-position-gluten-free-claims.html#a4.

CHAPTER 10

Supporting Food Health Claims in New Zealand's Regulatory Environment

CAROLYN E. LISTER

The New Zealand Institute for Plant and Food Research Limited,
Private Bag 4707, Christchurch, New Zealand
Email: carolyn.lister@plantandfood.co.nz

10.1 Introduction

In New Zealand and Australia, a new food standard (Standard 1.2.7[1]) to regulate nutrition content claims and health claims on food labels and in advertisements was introduced into law on 18 January 2013. Prior to this, although nutrient content claims were permitted, health claims were largely prohibited. The Standard was developed by Food Standards Australia New Zealand (FSANZ) as part of the Australia New Zealand Food Standards Code.[2] There was a three-year transition period but, since 18 January 2016, all health and nutrition claims must comply with Standard 1.2.7. It is important to note that, unlike some overseas jurisdictions, these regulations apply only to foods and supplemented foods. Dietary supplements and natural health products not in a food format or foods that make a therapeutic claim fall under different regulations (Table 10.1). Currently, dietary supplements have fewer restrictions on health claims, indeed, apart from restrictions on therapeutic claims, there is no claim framework for dietary supplements in New Zealand and Australia at present. A Natural Health Products Bill was under consideration that would have allowed health claims to be made only

Food Chemistry, Function and Analysis No. 22
Health Claims and Food Labelling
Edited by Siân Astley

Published by the Royal Society of Chemistry, www.rsc.org

Table 10.1 Regulations for claims applying to different categories of foods and dietary supplements.

Product format	Overarching legislation	Specific regulation for nutrition and health claims	Administration/ enforcement agency
Conventional food[a]	Australia New Zealand Food Standards Code[2]	FSANZ Standard 1.2.7 – Nutrition, Health and Related Claims[1]	Ministry for Primary Industries
Supplemented food[b]	NZ Food (Supplemented Food) Standard 2016[3]	FSANZ Standard 1.2.7 – Nutrition, Health and Related Claims	Ministry for Primary Industries
Dietary supplement[c]	Dietary Supplements Regulations 1985[4]	No separate specifications	Medsafe division of the Ministry of Health
Therapeutic[d]	Medicines Act 1981[5]	No separate specifications	Medsafe division of the Ministry of Health

[a]Each of the various Acts has a definition of food. These all have a similar effect and make the concept very broad, effectively covering anything that is intended or offered for human consumption. However, for the purposes of Standard 1.2.7 it excludes supplemented and medical foods plus dietary supplements.
[b]Supplemented food is a product that is represented as a food, but it has been modified in some way or had substances added to it so that it performs a physiological role. The supplementation or modification means it provides more than simple nutrition.
[c]Dietary supplements (including natural health products) are substance(s) for oral use that are packed in a controlled-dosage form and are intended to supplement the intake of substance(s) normally derived from food.
[d]Therapeutic means preventing, alleviating, treating, curing, or compensating for, a disease, ailment, defect, or injury; or influencing, inhibiting, or modifying a physiological process.

if included on a pre-approved list or able to be substantiated by the supplier. However, as of July 2018, this Bill has stalled, and because of this some companies have chosen to go down the dietary supplement route rather than foods because it is easier to make health claims.

Although food standards are developed by FSANZ, they are not responsible for enforcing them; this is the responsibility of the Australian States and Territories and New Zealand Agencies. In New Zealand, the Ministry for Primary Industries (MPI) is responsible for enforcing nutrition content and health claims. In addition to FSANZ standards, the New Zealand Food Act 2014[6] and Fair Trading Act 1986[7] also require that food labels and advertisements do not misinform consumers through false or misleading representations. Any breaches of these regulations may result in penalties, and changes in packaging and advertising will be required. In addition, such breaches have a significant effect on consumer trust, meaning it is important that the food industry is supported in making compliant health claims. Changes in regulations around food, generally, have been challenging for the food industry and, even five years after initial implementation, some food business operators (FBOs) are still struggling to comply with health claim legislation. FSANZ and MPI have been working closely with Australian departments in consultation with Australian and New Zealand food industry representatives to develop a guidance document.[8] As a research organisation, Plant & Food Research has also been working on finding ways to support the food industry. The following sections cover some of the areas that we have been working on and the approaches taken.

10.2 Navigating the Claim Options

A key challenge for FBOs has been to navigate the different claim options and determine what they can use for existing products or might target for new product development, as the types of claims and information needed to apply them is spread over different food standards and schedules. FSANZ Standard 1.2.7 covers various types of nutrition and health claims whilst others cover aspects such as nutrition information requirements (Standard 1.2.8[9]), and Schedules 1 and 4 contain recommended daily intakes and list pre-approved health claims, respectively.[10,11] The types of claims covered by Standard 1.2.7 are shown in Figure 10.1. Some foods are not permitted to carry claims, *e.g.* kava (*Piper methysticum*, a crop from the Pacific Islands), infant formulae, and foods that contain more than 1.15% alcohol by volume (other than energy content, carbohydrate content or gluten content; or salt or sodium content where not a beverage). There are also restrictions on

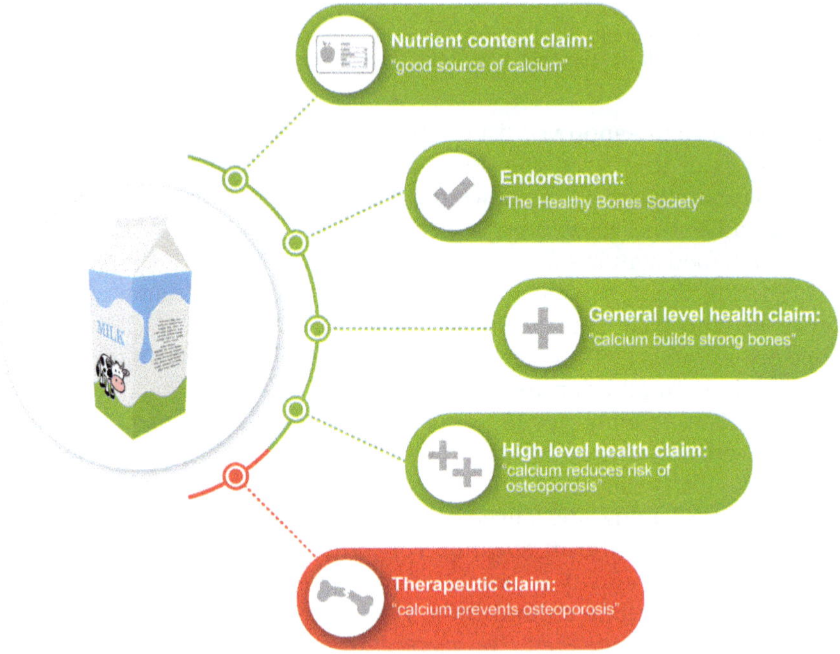

Figure 10.1 Examples of types of claims covered or prohibited by Australia New Zealand Food Standards Code: Standard 1.2.7 – Nutrition, health and related claims.

claims and representations about foods that are, or which contain, hemp products (covered in Standard 1.4.4[12]).

10.2.1 Nutrient Content Claims

Nutrient content claims tell consumers about the potential nutritional benefits of a food. They are important in the context of health claims, as in most cases they provide a framework for pre-approved health claims. Content claims can indicate the presence of a particular nutrient or the amount, *e.g.* "low in fat" or "good source of calcium". These claims need to meet criteria set out in Standard 1.2.7. Table 10.2 provides examples of the thresholds used for making different nutrient content claims for key nutrients (*e.g.* source or good source claims). For food products that contain nutrients or phytochemicals (*e.g.* anthocyanins, glucosinolates, lycopene, lutein) that do not have a recommended dietary intake (RDI), estimated safe and adequate daily dietary intakes (ESADDI) or other defined values in the Food Standards Code, only "presence" claims can be made (*i.e.* "contains"), *e.g.* "tomatoes contain lycopene" and not "tomatoes are a rich source of lycopene".

Currently, the FSANZ Code makes use of regulatory nutrient reference values (NRVs) for vitamins, minerals and protein, based on RDIs established in 1991[13] and, where such values were unavailable, the 1989 United States Food and Drug Administration (US FDA) ESADDI values.[14] NRVs for macronutrients and their components were drawn from other government recommendations. However, these NRVs are outdated and FSANZ is working on a proposal to review and update the values in the Code,[15] based on revised NRVs published in 2006 and an update on sodium and fluoride from 2017.[16] This review might shift thresholds for some nutrients as, currently, for example, vitamin C has a very low RDI compared with other jurisdictions.

10.2.2 Health Claims

Health claims refer to a relationship between a food or food component, and human health. Under FSANZ Standard 1.2.7, there are two classes of health claim:

- General level health claims, which refer to a nutrient or substance in a food and its effect on a health function (*e.g.* calcium, necessary for normal teeth and bone structure). These claims must not refer to a serious disease or a biomarker of a serious disease. In other jurisdictions, such claims might be referred to as structure–function claims.
- High-level health claims, which refer to a nutrient or substance in a food and its relationship to a serious disease or to a biomarker of a serious disease (*e.g.* calcium, enhances bone mineral density).

Health claims must be supported by sound scientific evidence, either pre-approved by FSANZ or self-substantiated. FBOs wanting to make general level

Table 10.2 Thresholds for making source, good source or excellent source claims on food labels. Daily intakes are based on an average adult diet of 8700 kJ and data are derived from FSANZ Standards 1.2.7[1] and 1.2.8[9] plus Schedules 1[10] and 4[11] of the Australia New Zealand Food Standards Code. For some nutrients there are additional requirements before claims can be made.

Nutrient	Adult RDI or ESADDI for labelling[a]	Threshold per serving	
		Source	Good source
Protein	50 g	5 g	10 g
Dietary fibre[b]	30 g	2 g	4 g
Omega 3 fatty acids	—	200 mg ALA or 30 mg total EPA and DHA[c]	60 mg total EPA and DHA[d]
Vitamins			
Biotin	30 μg[e]	3 μg	7.5 μg
Folate	200 μg	20 μg	50 μg
Niacin	10 mg	1 mg	2.5 mg
Pantothenic acid	5 mg[e]	0.5 mg	1.25 mg
Riboflavin (vitamin B2)	1.7 mg	0.17 mg	0.43 mg
Thiamine (vitamin B1)	1.1 mg	0.11 mg	0.28 mg
Vitamin A	750 μg	75 μg	188 μg
Vitamin B6	1.6 mg	0.16 mg	0.4 mg
Vitamin B12	2 μg	0.2 μg	0.5 μg
Vitamin C	40 mg	4 mg	10 mg
Vitamin D	10 μg	1 μg	2.5 μg
Vitamin E	10 mg	1 mg	2.5 mg
Vitamin K	80 μg[e]	8 μg	20 μg
Minerals			
Calcium	800 mg	80 mg	200 mg
Chromium	200 μg[e]	20 μg	50 μg
Copper	3 mg[e]	0.3 mg	0.75 mg
Iodine	150 μg	15 μg	37.5 μg
Iron	12 mg	1.2 mg	3 mg
Magnesium	320 mg	32 mg	80 mg
Manganese	5 mg[e]	0.5 mg	1.25 mg
Molybdenum	250 μg[e]	25 μg	62.5 μg
Phosphorus	1000 mg	100 mg	250 mg
Selenium	70 μg	7 μg	17.5 μg
Zinc	12 mg	1.2 mg	3 mg

[a]RDI = recommended dietary intake; ESADDI = estimated safe and adequate daily dietary intake.
[b]Dietary fibre can also make an excellent source claim at 7 g per serving.
[c]ALA = alpha-linolenic acid; EPA = eicosapentaenoic acid; DHA = docosahexaenoic acid.
[d]The food may contain less than 200 mg alpha-linolenic acid/serving.
[e]ESADDI (all others are RDIs).

health claims can base their claims on one of more than 200 pre-approved food–health relationships specified in Schedule 4[11] of the Food Standards Code or self-substantiate a food–health relationship (see Section 10.5). High-level health claims must be based on a food–health relationship that has been pre-approved by FSANZ; currently, there are 13 pre-approved high-level food–health claims.

Alternatively, an application can be made to add a new high-level food–health relationship to the pre-approved list.

In order to make health claims, foods must meet the nutrient profiling scoring Criteria (NPSC), as set out in Standard 1.2.7 and further detailed in Schedule 5.[17] The aim of NPSC scores is to ensure that health claims are not made on products that are high in energy, saturated fat, sugar or salt. FSANZ has developed an online calculator to help FBOs determine whether their product qualifies for a health claim.[18] If a food does not meet NPSC, it would be necessary to reformulate the product (*e.g.* lower salt, sugar and/or fat) in order to make a claim. FSANZ has also developed a user guide to assist FBOs with NPSC,[19] and some FBOs have reformulated products in order to apply health claims.

Unlike some overseas territories, FSANZ does not set the wording for nutrition and health claims, but there are guidelines that must be followed. In particular, it is important that the meaning and intent of pre-approved claims is not changed. General level claims can only refer to a "requirement for" or "contribution to" normal function. Terms like "inhibit", "reduce", "boost" and "increase", which imply inhibition or enhanced normal physiological function, must not be used. Similarly, words like "treatment of", "relief from", "prevention of", "medical", "therapeutic" and "symptoms" that suggest a therapeutic claim, cannot be used. These restrictions have presented some challenges to ensure FBOs and retailers do not exaggerate claims and wording complies with the Standard.

10.2.3 The New Zealand Food Composition Database – A Useful Tool to Support Claims

In order to make nutrient content claims and apply pre-approved health claims, it is critical that robust food composition data are available on which to base claims and/or determine nutrients that should be determined in a product. Plant & Food Research and the Ministry of Health jointly own the New Zealand Food Composition Database (NZFCD, www.foodcomposition.co.nz).[20] The database provides access to nutrient information for more than 2700 foods commonly eaten in New Zealand including fresh produce, raw ingredients, home-cooked recipes, and manufactured products specific to the New Zealand market. There is comprehensive information on 83 core nutrients, including carbohydrates, protein, fats, vitamins, minerals and dietary fibre, verified by independent laboratories. NZFCD includes all of the components for which nutrient content claims can be made.

Until recently, New Zealand food composition data was accessible in two ways: the Concise New Zealand Food Composition Tables or FOODfiles™. However, finding and printing data for a particular food or ingredient using these formats was not straightforward. Lacking wider context, food composition data delivered in these formats were just raw numbers and did not indicate directly what values might mean nutritionally or for labelling purposes. It was recognised that, to offer full value and exploit potential applications, the comprehensive food composition data needed to be easier to access and address the questions of users more readily. The website was upgraded and a search interface added so users can interrogate NZFCD more easily and outputs are more meaningful. For example, one of the new output formats is a "Potential Nutrient Claims Report" that identifies components in a food for which nutrient content claims can be made (Figure 10.2). These changes make it easier for the food industry to use nutrient content claims and identify which nutrients to focus on during development of new products.

Kiwifruit, Zespri Green (Hayward) Kiwifruit, Zespri, raw

FCDB food ID	L1026
Scientific name	*Actinidia deliciosa*
Food group	Fruit
Serving Size	74[a]
Recommended Serving Size	1 kiwifruit = 74 g

* Percentage daily intakes are based on an average adult diet of 8700 kJ as specified by FSANZ[b].

NUTRIENT	UNIT	QUANTITY PER SERVE	%DI* PER SERVE	QUANTITY PER 100 g	POTENTIAL CLAIM
NIP					
Energy, FSANZ	kJ	185	2 %	250	
Protein	g	0.9	2 %	1.2	
Fat, total	g	0.5	1 %	0.7	
Fat, saturated (SFA)	g	0.07	0 %	0.09	
Carbohydrate, available	g	6.7	2 %	9.1	
Sugars, total	g	6.5	7 %	8.8	
Dietary fibre	g	2.2	7 %	3.0	Source
Sodium	mg	2	0 %	2	
OTHER POTENTIAL CLAIMABLE					
Folate	µg	28	14 %	38	Source
Potassium	mg	222		301	Source
Vitamin C (ascorbic acid)	mg	63.0	157 %	85.1	Good Source
Vitamin K	µg	8.1	10 %	11.0	Source

[a] this value is adjustable on the web interface and values in table update if this is changed
[b] Food Standards Australia New Zealand

Figure 10.2 Example of a "Potential Nutrient Claims" report from the New Zealand Food Composition Database search tool.[20]

It also provides a framework to identify which of the pre-approved health claims might be applied, as long as the product meets NPSC requirements.

10.3 Antioxidant Claims

One of the areas that has caused issues since the implementation of Standard 1.2.7 has been antioxidant claims. Prior to implementation of the new standard, a number of FBOs had been making antioxidant claims. In many cases, use of the term "antioxidant" referred to the general mode of action of many nutrients and phytochemicals and antioxidant activity values, such as ORAC (oxygen radical absorbance capacity), were cited on labels. The challenge has been that antioxidants as a group have no officially recognised RDI or ESADDI and thus, under Standard 1.2.7, content claims cannot be made (*i.e.* cannot state "source of antioxidants" or "high in antioxidants"). However, claims can be made for specific nutrients that have pre-approved antioxidant/radical scavenging function effects, such as:

- Vitamin C contributes to cell protection from free radical damage
- Vitamin E contributes to cell protection from free radical damage
- Copper contributes to cell protection from free radical damage
- Manganese contributes to cell protection from free radical damage
- Selenium is necessary for cell protection from some types of free radical damage
- Zinc contributes to cell protection from free radicals.

Because the exact wording is not prescribed by FSANZ, these can be translated to antioxidant claims, as long as the intent of the wording is not changed and reference is always made to the specific nutrient. For example, it is possible to claim "blackcurrants are a good source of the radical scavenger vitamin C" but not that "blackcurrants are a good source of antioxidants". However, phytochemicals, commonly cited in the scientific literature as antioxidants, are not on the list of compounds with pre-approved health claims.

If antioxidant activity (*e.g.* ORAC value) is stated, any wording would be regarded as a general level health claim rather than a content claim and no such pre-approved health claims exist in Schedule 4. If general antioxidant claims are to be made, these will need to be characterised further before forming part of the food–health relationship. It would be necessary to define the food component(s) with antioxidant properties and develop a self-substantiation dossier. However, this might be challenging, as a suite of claims for antioxidants were rejected after the European Food Safety Authority (EFSA) found they could not be substantiated and it is likely that FSANZ would take a similar position. Currently, MPI directs FBOs to documents developed by the EFSA Panel on Dietetic Products, Nutrition and Allergies (NDA), where there are specific guidelines covering antioxidants that would be relevant for FBOs wanting to make general antioxidant claims.[21]

10.4 Developing Branded Products: vital**vegetables**®

One of the aims of Standard 1.2.7 was to enable the food industry to take advantage of market opportunities and provide opportunities for innovation. The **vital**vegetables® range is an example of such innovation with a strong focus on health.[22] The brand was developed as the result of an extensive research programme with funding from the vegetable industry and government sources. The project began before new health claims legislation was put in place, and the original plan was to focus on phytochemicals within single vegetable products, such as high glucosinolate broccoli and high lycopene tomatoes. However, under the new standard, it became apparent that this was going to be difficult, as it would require development of self-substantiation dossiers and significant investment in clinical trials. Instead, focus shifted to products containing a mixture of vegetable ingredients that were developed to utilise pre-approved health claims.

In the development of these fresh vegetable products, Plant & Food Research supported its industry partner through key considerations to deliver initial concepts to market. These included:

1. Product format: It was important to begin with a clear understanding of the food product. This was achieved by defining the product clearly in terms of types of ingredients, target consumers and end use (*e.g.* raw or cooked). For the initial release of **vital**vegetables®, product formats were salads, slaws and a vegetable medley.
2. Branding and health claim messaging: Consumer research was undertaken to understand what types of claims resonated within the confines of what is permitted by the Standard. Once the relevant health areas were identified, it was possible to map which nutrients had pre-approved health claims to fit with these. One of the key learnings from the commercialisation phase was the need for visibility at shelf level, which was achieved through a strong **vital**vegetables® branded range. Due to the apparent strength of the word vital, its use was extended to sub-brands that link products across the range through health benefits/claims. For the initial products released on the market, the following were selected:
    - **vital**immunity®: a good source of the antioxidant vitamin C and vitamin A to help maintain a healthy immune system.
    - **vital**bones®: designed with your bone health in mind; vitamin K and vitamin C to support bone structure.
    - **vital**sight®: designed with a focus on your vision; naturally high in provitamin A carotenoids to support healthy vision.
    - **vital**heart®: designed with your heart in mind; more than 10% of the RDI for vitamin C and contains phenolics.
3. Selection of ingredients and formulation: A key step in development was identifying which vegetable ingredients contained nutrients above the threshold to make nutrient content claims. Then, it was necessary to bring these ingredients together in different combinations to meet

consistently the nutrient thresholds all year round. Achieving the NPSC is easy for vegetable products but, because some of the products included dressings, this required careful consideration of dressing formulations to retain the claims.

4. Other consumer attributes: In development of these products, it was essential that ingredients were selected not only for their naturally high nutrient levels but also their taste and visual appeal. This was achieved through inclusion of more colourful components, such as purple cauliflower in the vegetable medley.
5. Nutrient quality assurance (NQA): Although not a requirement of the Standard, a NQA protocol was implement to provide assurances that the nutrient content at the end of shelf life is above the target levels required for the claim(s). The key nutrients on which claims are based are measured quarterly and a process for handling product failures was developed.

In addition, there were a number of other considerations during product development including packaging and recommendations on preparation (*e.g.* how to cook the medley to maximise retention of nutrients). Bringing these together achieved the goal of the project, which was to deliver a product with health benefits understood by consumers, that looks and tastes good, is locally grown, has an acceptable premium, and is convenient and simple to use.

10.5 Developing a Self-substantiation Dossier

Standard 1.2.7 differs from other international claim regulations in that it has provision for the food industry to self-substantiate a food–health relationship in order to make a general level health claim. The intent of including self-substantiation within the Standard was to support innovation, allowing FBOs to take advantage of market opportunities and maximise the export potential of value-added products. However, to make a new claim, it is necessary to demonstrate that, on the basis of evidence from a systematic review, there is a cause-and-effect relationship between the food or property of a food and the health effect claimed. The key objective of the substantiation process is to determine whether evidence for this relationship is sufficient and certain that it "is unlikely to be reversed by an additional well-conducted, high-quality study".

Developing a self-substantiation dossier is a significant investment and must be done appropriately to meet the requirements of the Standard. Systematic reviews comprise a clearly formulated question and use of systematic and explicit methods to identify, select, critically appraise, extract and analyse data from relevant research.[23] For substantiation of food–health relationships under the Standard, relevant research must include human studies using experimental or observational designs. Integral to the process are a systematic approach, transparency, comprehensiveness, evidence in humans and causality.[24,25]

Both FSANZ and MPI have developed guidance documents on establishing food–health relationships for general level claims.[25,26] These provide lots of useful information to guide FBOs through the process. In addition, the New Zealand Institute for Plant & Food Research has built on this guidance to help steer companies through the process. Together with Drummond Food Science Advisory, Plant and Food Research developed a systematic review for Zespri®, which supported the claim that "Zespri® Green Kiwifruit can contribute to normal bowel function". In 2014, this claim was the first lodged with FSANZ, from New Zealand, under the new regulations. Plant & Food Research continues to work with FBOs to build health claim dossiers, but success is challenging.

A key part of the self-substantiate review is assessment of study quality. Schedule 6 lists six required elements with regard to the quality of clinical studies, as a minimum, namely a clearly stated hypothesis, minimisation of bias, adequate control for confounding factors, consideration of participants' background diets and other relevant lifestyle factors, study duration, adequate follow-up to demonstrate the health effect, and statistical power.[27] Use of a study appraisal tool can support critical assessment of individual clinical studies and help identify studies that are more likely to have generated meaningful, high-quality data. In the FSANZ guidance document,[25] six different tools are suggested for assessing study quality. However, none of these assesses all of the elements of study quality that are required for self-substantiation. Plant & Food Research has been developing a quality appraisal tool for intervention studies to assist the food industry in developing dossiers (Table 10.3), based on a Health Canada tool[28] that has been adapted to include elements required by FSANZ, such as study hypothesis, consideration of participant's background diet and other lifestyle variables, study duration and statistical power.

The disciplines of systematic review and assessment of study quality have highlighted many issues with clinical trials from design to reporting. Admittedly, the majority of trials were not designed with the intention of making health claims for food products and the underlying scientific questions are often quite different from those asked by regulators. The benefits of conducting such reviews, however, have enabled these differences to be considered carefully, as FBOs and researchers move towards future studies, creating a stronger framework for other self-substantiated claims. In addition to the quality appraisal tool, Plant & Food Research has been developing a clinical trial toolkit to improve the design of trials, especially for food-related claims, to help ensure compliance with regulatory requirements. Hopefully, going forward, this will increase the chances of success for FBOs wanting to invest in a self-substantiation dossier and an approved health claim.

10.6 Conclusions

Making health claims in New Zealand's regulatory environment presents challenges for the food industry, but considerable progress has been made since implementation of the new standard (Standard 1.2.7) in 2013. Making a successful claim requires understanding of three key elements: regulations,

Table 10.3 Quality appraisal table developed for assessment of study quality of clinical trials for self-substantiation dossiers (adapted from Health Canada 2009[28]).

Area	Question	Score[a]
Hypothesis	Was there a clearly stated and appropriate study hypothesis?	
Background information on participants	Were the participants selected the intended target group for the food–health relationship?	
	Were inclusion and/or exclusion criteria reported (*e.g.* age, gender, disease status, medication)?	
	Were background diet and other relevant lifestyle variables considered and reported?	
Group allocation	Was the study reported to be randomised?	
	Was the randomisation method reported?	
	Was the randomisation method appropriate?[b]	
	Was allocation concealed?[c]	
Blinding	Were study participants blinded to the intervention received?	
	Were research personnel blinded to the intervention received by the participants?	
Study design	Was the duration of the intervention (and if relevant follow-up) appropriate for the health effect?	
	In crossover studies was there an adequate washout in between study arms?[d]	
	Were the participants in different treatment groups compared at baseline?	
Attrition	Was attrition numerically recorded?	
	Were the reasons for dropouts provided?[e]	
Exposure/ intervention	Was the type of food intervention and consumption method fully described (*e.g.* variety, composition, matrix)?	
	Was the amount of food described (serving size and frequency of consumption)?	
Health effect	Was the methodology used to measure the health effect sufficiently reported?	
Statistical analysis	Was the sample size sufficient to determine statistical significance?	
	Was a between-group statistical analysis of the health effect conducted (*i.e.* control *vs* intervention)?	
	Was an intention-to-treat analysis conducted?[f]	
Potential confounders	Were potential confounders of the food–health relationship considered?[g]	
TOTAL SCORE (max 22)		
High quality (score 16–22)		
Medium quality (score 9–15)		
Low quality (score 0–8)		

[a]Scoring: Yes = 1, No or not reported = 0

[b]Examples of appropriate randomisation include the use of computer-generated random number table, while date of birth and alternate allocation are examples of inappropriate methods of randomisation.

[c]Allocation concealment refers to the method used to implement the random allocation sequence (*e.g.* numbered envelopes containing the assignment). It protects the assignment sequence before and until allocation.

[d]For trials with a parallel design this question is not relevant and should be scored as a YES so as to not unfairly lose a point.

[e]If there was no participant attrition then reasons for withdrawals/dropouts are not relevant and should be scored as a YES so as to not unfairly lose a point.

[f]If there was no subject attrition, a per-protocol analysis is appropriate and an intention-to-treat analysis not applicable. In such a case check YES so as to not unfairly lose a point.

[g]Confounding could have occurred during subject selection, study conduct or data analysis. If randomisation is successful and between groups differences that may have occurred during the study are considered during statistical analysis, then confounders were considered.

science and consumers (including marketing). The food industry is usually closest to the consumer, but the interface between regulations and science can present challenges for FBOs. As a research organisation, Plant & Food Research has been working together with the New Zealand enforcement agency (MPI) to support the industry in this area, and past work supporting the food industry has led to the successful launch of products on the New Zealand market with health claims. For those FBOs wanting to make health claims, there are some key considerations:

- Invest wisely and design clinical trials carefully, taking into account regulatory considerations.
- Focus on areas where there is a significant body of high-quality research to support new claims.
- Potential claims need to be considered at the outset of product planning.
- Understand the consumer, conduct market research about their needs and build consumer trust and understanding.
- The self-substantiation route is not easy; it can be done but understand clearly what is allowed under the regulations.
- Utilisation of pre-approved claims requires a good understanding of food composition.

Acknowledgements

The New Zealand Food Composition Database is funded by the New Zealand Ministry of Health and additional funding to provide new search functionality was provided through the Ministry of Business, Innovation & Employment (MBIE) Strategic Science Investment Fund (SSIF).

The **vital**vegetables® research programme was funded by Horticulture New Zealand, Plant & Food Research and the Department of Environment and Primary Industries Victoria, with Horticulture Australia Limited matched funds from the Australian Government and support from the New Zealand vegetable sector.

References

1. Australia New Zealand Food Standards Code – Standard 1. 2.7 – Nutrition, health and related claims. https://www.legislation.gov.au/Series/F2015L00394 (accessed July 2018).
2. Australia New Zealand Food Standards Code. http://www.foodstandards.govt.nz/code (accessed July 2018).
3. NZ Food (Supplemented Food) Standard 2016. https://www.mpi.govt.nz/food-safety/labelling-and-composition/applying-labelling-and-composition-requirements/supplemented-food/ (accessed July 2018).
4. Dietary Supplements Regulations 1985 (SR 1985/208). http://www.legislation.govt.nz/regulation/public/1985/0208/latest/DLM102109.html (accessed July 2018).

5. Medicines Act 1981. http://www.legislation.govt.nz/act/public/1981/0118/69.0/DLM53790.html (accessed July 2018).
6. Food Act 2014. http://www.legislation.govt.nz/act/public/2014/0032/latest/DLM2995811.html (accessed July 2018).
7. Fair Trading Act 1986. http://www.legislation.govt.nz/act/public/1986/0121/64.0/DLM96439.html (accessed July 2018).
8. Getting your claims right: A guide to complying with the Nutrition, Health and Related Claims Standard of the Australia New Zealand Food Standards Code. http://www.foodstandards.govt.nz/publications/Pages/gettingyourclaimsright.aspx (accessed July 2018).
9. Australia New Zealand Food Standards Code – Standard 1.2.8 – Nutrition information requirements. https://www.legislation.gov.au/Series/F2015L00395 (accessed July 2018).
10. Reference: Australia New Zealand Food Standards Code – Schedule 1 – RDIs and ESADDIs. Available at: https://www.legislation.gov.au/Series/F2015L00491 (accessed July 2018).
11. Australia New Zealand Food Standards Code – Schedule 4 – Nutrition, health and related claims. https://www.legislation.gov.au/Series/F2015L00474 (accessed July 2018).
12. Australia New Zealand Food Standards Code – Standard 1.4.4 – Prohibited and restricted plants and fungi. https://www.legislation.gov.au/Series/F2015L00416 (accessed July 2018).
13. Recommended Dietary Intakes for Use in Australia. http://www.nhmrc.gov.au/_files_nhmrc/file/publications/synopses/withdrawn/n6.pdf (accessed July 2018).
14. Food and Nutrition Board: Institute of Medicine, Recommended Dietary Allowances, National Academy Press, Washington, D.C., 10th edn, 1989.
15. P1047 – Review of regulatory nutrient reference values. http://www.foodstandards.gov.au/code/proposals/Pages/P1047.aspx (accessed July 2018).
16. Nutrient Reference Values for Australia and New Zealand (including Recommended Dietary Intakes). https://www.nrv.gov.au/resources (accessed July 2018).
17. Australia New Zealand Food Standards Code – Schedule 5 – Nutrient profiling scoring method. https://www.legislation.gov.au/Series/F2015L00475 (accessed July 2018).
18. Nutrient Profiling Scoring Calculator for Standard 1.2.7. http://www.foodstandards.gov.au/industry/labelling/pages/nutrientprofilingcalculator/default.aspx (accessed July 2018).
19. Short guide for industry to the Nutrient Profiling Scoring Criterion in Standard 1.2.7 – Nutrition, health and related Claims. http://www.foodstandards.govt.nz/industry/labelling/Pages/Short-guide-for-industry-to-the-NPSC.aspx (accessed July 2018).
20. New Zealand Food Composition Data. http://foodcomposition.co.nz/ (accessed July 2018).

21. EFSA Panel on Dietetic Products, Nutrition and Allergies (NDA), *EFSA J.*, 2011, 9(12): 2474.
22. vitalvegetables® http://vitalvegetables.com/ (accessed July 2018).
23. Cochrane Handbook for Systematic Reviews of Interventions. http://www.cochrane.org/training/cochrane-handbook (accessed July 2018).
24. Guidance Document for preparing a submission for food health claims using an existing systematic review. https://www.canada.ca/en/health-canada/services/food-nutrition/legislation-guidelines/guidance-documents/guidance-document-preparing-submission-food-health-claims-2011.html (accessed July 2018).
25. Guidance on establishing food-health relationships for general level health claims (as described in Schedule 6 of Standard 1.2.7 – Nutrition, Health and Related Claims). http://www.foodstandards.govt.nz/publications/Pages/Guidance-on-establishing-food-health-relationships-for-general-level-health-claims.aspx (accessed July 2018).
26. Systematic review of a food-health relationship – Guidance document on how to self-substantiate a food-health relationship in order to make a new general level health claim. https://www.mpi.govt.nz/food-safety/labelling-and-composition/health-claims-for-high-value-foods/self-substantiation-of-health-claims/ (accessed July 2018).
27. Australia New Zealand Food Standards Code – Schedule 6 – Required elements of a systematic review. https://www.legislation.gov.au/Series/F2015L00476 (accessed July 2018).
28. Guidance Document for Preparing a Submission for Food Health Claims. https://www.canada.ca/en/health-canada/services/food-nutrition/legislation-guidelines/guidance-documents/guidance-document-preparing-submission-food-health-claims-2009-1.html (accessed July 2018).

CHAPTER 11

Latin American Food Labelling: Review and Recent Developments

MARCELA REYES* AND CAMILA CORVALAN

Institute of Nutrition and Food Technology (INTA), University of Chile, Chile
*Email: mreyes@inta.uchile.cl

11.1 Background

Health problems associated with all forms of malnutrition are the main cause of death and loss of healthy years of life around the world.[1–3] It is estimated than 158 million children under 5 years of age are stunted whilst excess weight and obesity affect 42 million children and 1.9 billion adults.[1–3]

Nutritional problems are not evenly distributed within the Latin American region. Stunting amongst children under 5 years of age is as low as 3–6% in countries like Chile, Uruguay and Brazil, but almost half of under-5s are malnourished in Guatemala. Countries like Colombia,[4] Bolivia[5] and Peru[6] have an intermediate prevalence, between 15 and 25%.[7] In contrast, overweight and obesity amongst under-5s is less than 5% in countries like Colombia[4] and Guatemala,[8] but around 35% in Chile; the prevalence of overweight and obesity in Peru,[6] Uruguay[9] and Brazil[10] is between 10 and 20%.[7] Amongst adults, overweight and obesity has become a more homogenous problem, with 40–60% of adults (or women alone, depending on the statistics available) being overweight or obese.[4–6,8–14] However, chronic undernutrition among adults is highly variable in these countries; from less

Food Chemistry, Function and Analysis No. 22
Health Claims and Food Labelling
Edited by Siân Astley

Published by the Royal Society of Chemistry, www.rsc.org

than 3% in Uruguay[9] and Chile[13] to 35% among Peruvian[6] women and 29% among Guatemalan women.[8]

In all Latin American countries where undernutrition is an important health problem, the greatest burden is seen amongst the poorest and least educated populations. However, disparities in weight vary from country to country. In Chile,[13] Uruguay,[9] Brazil,[10] Colombia[4] and Ecuador,[11] overweight is higher amongst populations with low socio-economic status (SES) or educational achievements but, in countries like Bolivia, Guatemala and Mexico, the differences are reversed (*i.e.* greater overweight among the richest populations). There are also countries with mixed disparity, with the highest rates of overweight among adults from low SES and children from higher SES (*e.g.* Argentina[14] and Peru[6]).

Obesity and non-communicable diseases (NCDs) are related to diets with excess energy, sugars, saturated or trans fats, and sodium. The 2016 Global Burden of Disease study indicated that poor dietary habits were the second greatest risk factor for mortality globally, representing 15% or almost one in five deaths.[15] In fact, the role of a healthy diet (and physical activity) in preventing illness was highlighted 15 years ago by the World Health Organization (WHO) Expert Committee report *Diet, Nutrition and the Prevention of Chronic Diseases*,[16,17] which recommended intakes of key nutrients to prevent development of NCDs.

The combined problem of obesity and NCDs has been approached using a variety of actions, largely targeted at individuals (compared to a population-wide approach) and, usually, focused on therapeutic rather than preventative actions. These measures have not, however, been able to address the continued increase in prevalence of such health problems. Thus, increasingly, a new perspective is evolving, focusing on environmental aspects that shape diet beyond the individual, *i.e.* sociocultural, physical and economic aspects that influence a population's diet, collectively known as the food environment.[18] Changes at the level of food quality, labelling and marketing, cost, retail (what is being sold where) and availability (food provided or sold at school or elsewhere) impact nutrition and health, beyond individual factors such as knowledge, attitudes or capacity for change.

Currently, food environments in big cities and some countries are obesogenic, increasing the chances of access to energy-dense foods, with high content of nutrients of concern (*i.e.* sugars, fats, sodium), heavily marketed/advertised, affordable/cheap, and labelled with cryptic nutritional information.[19–21] A few decades ago, this was true only for high-income countries, but now it is also true amongst low- and middle-income countries. Even when, compared with other regions of the world, such as Europe and North America, traditional cooking and consumption of minimally-processed foods predominate,[22–24] an important dietary shift has occurred in Latin American diets in recent decades. In fact, the commercialisation of highly processed products (*i.e.* ultra-processed food products [UPF]) is increasing rapidly across the region, according to the Pan American Health Organization [PAHO] 2015 report.[25] This and other studies have analysed

UPF sales (Euromonitor) from retail stores and fast-food outlets (2000–13) in 13 countries of the region. Purchase of UPF products (*i.e.* those formulated and manufactured industrially, employing processes and ingredients not commonly used in traditional culinary preparations)[26] has been linked to unhealthy diets (*i.e.* low intakes of proteins, fibre and micronutrients, and high intakes of sugars, fats and energy),[22,27,28] weight gain, development of hypertension, and breast cancer.[25,29–31] The PAHO report showed UPF sales increased in every country, except Argentina and Venezuela, during local financial crises. The report also showed sales of such products are linked to increased urbanisation and income. Latin America is fourth with respect to UPF product intakes, reaching 130 kg per capita in 2013, after 26% growth over a 13-year period; Bolivia, Uruguay and Chile have still greater growth in this sector (over 50%).[25]

Thus, most Latin American countries, albeit to different extents, are introducing more foods with greater degrees of processing. Besides the nutritional content of those foods, these products are usually heavily marketed/advertised, affordable/cheap and socially valued, characteristics that are shifting food environments to less healthy ones.[26] According to the new paradigm, which highlights structural measures – aimed at modifying food systems or food environments rather than an individual's knowledge/attitudes/capacities – as more cost-effective in reducing the impact of obesity and NCD,[18] several countries in the region have implemented or are on their way to implementing actions to make food environments healthier. Examples include Mexican taxes on sugar-sweetened beverages and non-essential energy-dense foods,[32,33] the Brazilian initiative to include 30% of family farmer supplies in school food programmes,[34] and Chilean restrictions on food marketing targeting children as well as the mandatory use of warning labels for unhealthy foods.[35]

Improved food labelling is one aspect of strategies aimed at shaping healthier food environments, impacting healthier purchases and, ultimately, improving diets at the population level.[36,37] Thus, nutritional labelling on food packaging can be considered a tool for implementing policies, through which governments, the food industry and health-related non-governmental organisations (NGOs) might re-orientate food selection. In fact, the WHO considers food labelling an important means of providing consumers with accurate, standardised and, most importantly, understandable information about the nutritional content of products, thus informing healthier decisions.[16,17] Other international agencies and NGOs, like UNICEF and the World Cancer Research Fund, also agree that food labelling is an opportunity for improving the food environment and diet of a population.[17,38] From a health perspective, nutritional labelling on foods aims to guide food selection towards healthier options, and is an important source of information for consumers when included as part of the strategies facilitating such options.[36] Food labelling should also improve understanding of the nutritional quality of a product, enabling informed and interested consumers to compare products and choose the best option. Secondly, nutritional labelling can also be a stimulus for the food industry to improve the quality/reformulate products.[39–41]

At the global level, food labelling characteristics are regulated by *Codex Alimentarius*, a programme created by the Food and Agriculture Organization (FAO) and the WHO, establishing guidelines for different aspects related to food. According to *Codex Alimentarius*, food labelling is a "description that has the intention to inform consumers about the nutritional properties of a food".[42] Most countries in the Latin American region take into account these guidelines, as related to nutritional declarations (*i.e.* standardised information about the nutritional content of a food). In general, Latin American countries have policies for nutritional labelling, but what these include differ from country to country. Some countries have common regulations, as in the case of Central America (*i.e.* Guatemala, El Salvador, Nicaragua, Honduras and Costa Rica)[43] or countries from Mercosur (*i.e.* Brazil, Argentina, Paraguay and Uruguay).[44] With the exception of Peru[45] and Colombia,[46] in most countries nutrient declarations are mandatory, the only variation being the nutrients that must be included. For instance, in Central America declarations must include energy, total and saturated fats, carbohydrates, proteins and sodium, whereas Mercosur countries also include fibre and trans fats. Chile[47] and Mexico[48] require sugars to be included (as grams or percentages of carbohydrates). Generally, most countries accept nutrient declarations as per 100 g or per serving but, in Chile, both (per 100 g and per serving) are required on the labels. Most countries also ask for a list of ingredients, but these do not include a quantitative declaration (although in Chile amounts of non-caloric sweeteners must be declared when present). Colombia[46] only requires a nutrient declaration when a health claim is made but, in Peru, there is no such requirement.[45]

Although food labelling is a tool for improving diets, most people do not read or understand the information provided, even amongst highly educated populations. Studies suggest that understanding of traditional nutritional labelling is around 30%.[49,50] Thus, in line with calls from international health agencies, supplementary information is being included. This approach aims to provide simpler and more easily understood nutritional information. According to *Codex Alimentarius*, supplementary nutritional information is part of food labelling and defined as "information presented in a wide variety of formats, having the intention to increase the understanding of the nutritional value of a food by consumers and helping in interpreting the nutrient declaration".[51] Supplementary information includes health claims and other simpler or summarised ways of presenting traditional nutritional information. According to *Codex Alimentarius*, a "claim is any representation which states, suggest or implies that a food has particular characteristics relating to its origin, nutritional properties, nature, production, processing, composition or any other quality".[42] In that context, claims are understood as marketing strategies that can be substantiated by manufacturers and, thus, most countries in the region clearly list messages that can be used as health claims and their corresponding nutritional criteria (*i.e.* Chile,[52] Colombia,[46] Mexico,[48] countries from Central America and Mercosur[44]). Most countries (*i.e.* Chile,[52] Colombia,[46] Ecuador,[53] Mexico,[48] countries from Central America and Mercosur[44]) also mandate that, when a health claim is nutrient-related, a declaration for that nutrient must be included (*i.e.* declaration of calcium content if any health claim relating to calcium is made).

For increased visualisation, claims and symbols are commonly on the front of packaging and collectively termed front-of-package (FOP) labelling. Over the last 25 years, several countries have proposed voluntary use of FOP labels, which has increased rapidly around the globe.[36,54,55] Overall, FOP labels have improved visualisation/awareness and understanding of nutritional labelling, compared with traditional approaches.[56–62] However, the impact on improved purchasing habits and the healthiness of choices varies according to design and ways in which information is displayed.[63–67]

11.2 FOP Labelling in Latin American Countries

Experience with FOP can be tracked from several decades ago and did not started in Latin American countries: Sweden was the first country (in the late 1980s) to use FOP labels to inform consumers about the nutritional content of packaged foods. They implemented a simple green (or black) logo with a white keyhole when specific nutrition criteria were achieved. The logo is a global or summary FOP label (*i.e.* provides a global qualification) and is meant to tag healthier foods (*e.g.* whole grains). It is still being used voluntarily in Sweden as well as other Nordic countries, and is known as the Keyhole or Nordic Logo.[68–70] FOP labels based on the same framework (voluntary summary positive logos) have been implemented in several countries globally (mainly in Europe, but also Asia, Africa and the Americas). The majority were developed or supported by the Choices Programme (www.choicesprogramme.org), which was created in the Netherlands in 2006, with help from researchers in the nutrition field,[37] after a government call to the food industry. Similar initiatives have been supported by NGOs like the American Heart Association.[70] In the late 1990s, guideline daily amounts (GDAs) were developed in the UK. These FOP labels comprise several coloured ovals, indicating amounts of each nutrient of interest in a serving as well as the percentage that quantities represent of the daily recommendation for that nutrient.[71,72] During the 2000s, the food industry implemented this approach voluntarily.[73,74] However, in 2007, a second type of labelling was developed and implemented in the UK, based on traffic light colour coding for the amounts of different nutrients of interest.[75,76]

European FOP labels are good examples of the wide spectra of characteristics that FOP labels may have. These can be categorised as providing summary or global information (*e.g.* keyhole) or nutrient- or ingredient-specific information (*e.g.* traffic light system and GDAs). They can also be grouped according to their focus on positive (*e.g.* keyhole) or negative messages (*e.g.* red in the traffic light system). An additional FOP classification has been proposed, based on users' perspectives, *i.e.* degree of additional understanding needed by consumers to make a healthy choice. Hodgkins *et al.*[77] classified FOP labels as non-directive (numeric data or detailed text, which need be de-codified and interpreted by user, *e.g.* GDAs), semi-directive (include detailed data, but are simpler to understand; *e.g.* traffic light system) or directive (communicate a global decision, without detailed data, *i.e.* decision has been made for the user, *e.g.* keyhole).

Generally speaking, compared to standard nutritional information, FOP labels have been shown to perform better with respect to visualisation, use, understanding and healthier food selection among adults and children.[78–80] However, performance varies across different types of FOP label.[63,64,81] Based on neurobiological characteristics of humans, some aspects of FOP labels have been identified as potentially more effective when delivering a message that will shape choice and, ultimately, nutrition and health. It is important to take into account that – despite general self-awareness – decisions about food selection are not entirely rational. This can be explained by several factors, such as priming to enjoy food. Also, food cues (looking, smelling, touching or tasting) activate reward systems in the brain,[82,83] which, in turn, are perceived as increased appetite and promote feeding actions.[84]

Another aspect to be considered is the human tendency for saving cognitive effort in frequent decision-making processes, especially everyday decisions, like those related to diet.[85] Such decisions go *via* a neurological shortcut that takes little account of rational thinking and is based mainly on simple and visual messages. Taking these factors into account, the influence of marketing strategies in food selection are understandable;[66,86–89] in fact, evidence suggests these messages activate brain reward systems directly, at least when appealing to hedonic elements.[90] However, it does also mean that simple/directive nutritional messages can also act through the same system.[90] Thus, even when people value nutritional information,[56,58,60,77,91] the more directive a FOP label is, the more effective it will be in affecting healthier food choices.[92–97] Besides simplicity,[36] there are other elements in design that are related to improved visualisation, which is the first obstacle to overcome in delivering nutritional messages. The use of pictorial elements and high contrast colour, light and shape are key.[98] Colour also influences perception, with black, grey and red inducing feelings of "unhealthiness", even without text.[99] Similarly, shape also affects response, and symbols associated with traffic (*i.e.* stop-signs or triangles) discourage purchase/consumption.[99]

11.2.1 Mandatory FOP in Latin American Region: Cases of Mexico, Ecuador and Chile

Across Latin America, a peculiarity regarding food labelling is the fact that, over the last five years, Mexico, Ecuador and Chile have implemented mandatory FOP labelling. Moreover, in these countries, three completely different labelling systems have been implemented: non-directive GDA in Mexico (2015), semi-directive traffic light system in Ecuador (2015) and directive warning labels in Chile (2016).

In the case of Mexico, in 2014, the Ministry of Health and Finance changed nutrient labelling regulations to make the use of FOP information, based on GDA, mandatory from June 2015 with a one-year implementation period. This was part of the Mexican national strategy to prevent and control overweight and diabetes.[100] Label specifications depend on whether the food is packaged individually (*i.e.* pack content less than a standard serving) or in a family pack (*i.e.* content equal or greater than the standard serving) (see Figure 11.1). The general idea was to provide information on the amounts of saturated and other fats, and sugars (all presented as calories), sodium (milligrams) and energy (or per serving for family packs). The information is also presented as percentage of recommended daily intakes. Family packs, which report energy content and nutrients based on servings, must also report the (expected) number of servings in the packet. The labels must be displayed FOP for those with a surface area greater than 60 cm^2; these can be shortened to only energy in cases of smaller surfaces. Details including technical and graphic specifications are publicly available elsewhere.[101]

To date, there has been no evaluation of impact for this regulatory measure, but there are reports raising concerns about the use[102] and potential effects of such labels, given 2015 data suggesting that only around 12% of the population knew how many calories they should consume a day and would, therefore, have limited understanding of GDA-based FOP labels.[103] Moreover, less than one-third of university students of nutrition could interpret GDA correctly, and only 2% answered correctly regarding amounts of each product they could eat without exceeding the recommended intake.[104]

Ecuador started work on a new FOP label to encourage healthier food decisions in 2012, convening different interested actors for the discussion. A traffic light system was selected, using colour coding and short phrases to deliver information about total fats, sugars and salt contents (Figure 11.2).[53,105] The label must have a white or grey background (depending on the dominant pack colour) and must be of a size proportional to the primary information on the package (*i.e.* name and other advertising), where the overall size is greater than 19.4 cm^2. It is worth noting that the label can be used on the main (front) or secondary (back) phases of the packet and so it is not strictly FOP labelling. Nutrient profiling supporting different contents (*i.e.* high, medium or low) are based on PAHO *Recommendations on the Marketing of Food and Non-Alcoholic Beverages to Children in the Americas*, which were based on UK traffic light nutrient profiling.[106] There has been some evaluation of this legislation that suggest the labels are recognised widely by the public, who also value the relevance and usefulness, but it is not clear whether these labels have changed dietary habits.[105,107]

In Chile, after five years of discussion in Parliament, the Food Labelling and Marketing Law was approved in 2012, requiring foods with high energy, sodium, saturated fats or sugars contents to have a FOP warning label clearly stating such. After three years of work and debate, a regulation was published

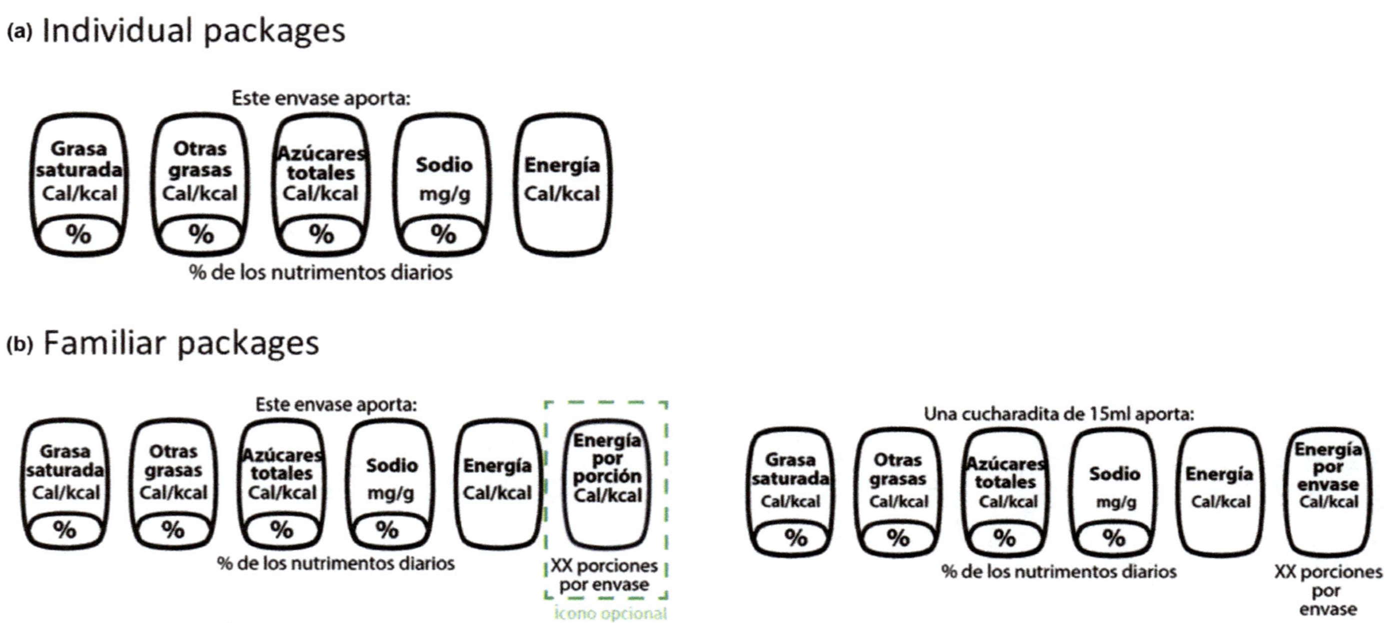

Figure 11.1 Mandatory FOP labelling implemented in Mexico (2015–16). (a) FOP label for individual packages, which reads: "This package provides"; (ovals from left to right) "Saturated fat [Cal/kcal and %], Other fats [Cal/kcal and %], Total sugars [Cal/kcal and %], Sodium [mg and %] and Energy [Cal/kcal]"; "% of daily recommendations". (b) FOP label options for familiar packages; both of them read: (ovals from left to right) "Saturated fat [Cal/kcal and %], Other fats [Cal/kcal and %], Total sugars [Cal/kcal and %], Sodium [mg and %] and Energy [Cal/kcal]"; "% of daily recommendations"; "Energy per serving [Cal/kcal], XX servings per package". Both options only differed in the first sentence, which is either "This package provides" or "A tablespoon of 15 mL (as example of serving) provides", and in the fact that the last oval is non-mandatory in the first option.

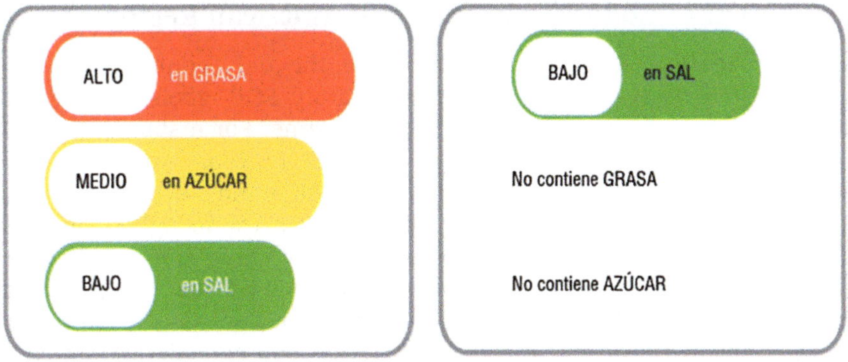

Figure 11.2 Mandatory FOP labelling implemented in Ecuador (2015). Left label reads: "HIGH in FAT, MEDIUM in SUGARS, LOW in SALT". Right label reads: "LOW in SALT, Has no FAT, Has no SUGAR".

Figure 11.3 Mandatory FOP label implemented in Chile (2016). Octagons read, from left to right: "HIGH IN SUGARS, HIGH IN SATURATED FATS, HIGH IN SODIUM, HIGH IN CALORIES"; all octagons display in the bottom: "Ministry of Health".

in 2015 outlining characteristics of these warning labels. Products using up to four black and white stop-signs with supporting text, indicating high in sugars, saturated fats, sodium or calories, started appearing in June 2016 (Figure 11.3).[108] The labels must be on the front of packets when the surface area is greater than 60 cm^2 and specific nutrient profiling was developed to inform this regulation.[35,47] Details of the implementation are available elsewhere.[108,109]

Chile was the first country to implement FOP warning labels and the labels are different from other examples, because they are nutrient-specific rather than a summary, and focus exclusively on negative aspects of the product as well as being directive, needing no further interpretation by users. Evaluations were done by the Chilean Ministry of Health after 6 and 18 months, enforcement effort and large food industry compliance[110,111] were examined and the results showed that 92.9% of the public recognised the labels and 43.8% said they used them to inform food purchases;[112] similar results were obtained among children in another report early in the implementation phase.[113] Qualitative data also suggest the new labels are known, understood and used for informing food purchases, highlighting the important role of young children as drivers of these changes.[114]

11.2.2 Other Countries in the Latin American Region

Although Mexico, Ecuador and Chile are the only countries to have implemented regulations regarding mandatory FOP labels, other countries in the region are working towards the same. For instance, Uruguay and Peru have approved FOP labels similar to those in Chile.[115,116] Currently, Brazil is debating which FOP label might be best, with different options being discussed, including GDA, traffic lights, only red traffic lights, and black and white triangles, as warning labels. Others are discussing whether they should move towards FOP labelling, as compared with more extensive non-directive information. Central American and Caribbean countries, and Colombia, are also considering which of these approaches should be implemented, when, and what elements should be mandatory.

FOP labelling systems in Ecuador and Chile are related to specific nutrient profiling criteria, so the food industry is clear about what must be included. It is not the purpose of this chapter to review the nutrient profiling system or its characteristics, but it is worth noting that most FOP labels, besides GDA, need such a system and the effectiveness of labelling policies are closely linked to the nutrient profiling scheme selected. Consequently, PAHO has convened an expert committee and proposed nutrient profiling to supporting labelling and other regulatory measures in the Americas.[117] Several countries are considering the nutritional criteria specified when implementing their mandatory FOP labels.

11.3 Conclusions

Latin America is leading the world in implementing directly mandatory, easy-to-understand nutrition labelling, usually on the front of packaging. This approach has been used voluntarily for more than 25 years in countries from other regions, but Mexico, Ecuador and Chile have opted to make such labelling mandatory. Other countries in the region are also advancing mandatory FOP labelling. Interest in FOP labels has increased globally, because of the potential to provide dietary information in a simple format and improve dietary choices around purchase and consumption as well as health. Small-scale experimental research suggests that simpler, more directive labels perform better in this respect, but there is little information regarding the success of large-scale, real-world experiences. In Latin America, three countries have chosen different approaches for FOP labels: directive (Chile), semi-directive (Ecuador) and non-directive (Mexico). Thus, over the next few years, Latin America will be able to inform other regions about the best (most effective) option to implement easy-to-understand mandatory labels and achieve improved population-wide dietary behaviours, tackling important health burdens (*i.e.* NCDs) derived from unhealthy intake (*i.e.* malnutrition).

References

1. F. Branca and A. Lartey, Ending malnutrition in all its forms? A decade of opportunity, 2016. http://www.who.int/mediacentre/commentaries/ending-malnutrition-opportunity/en/ (accessed March 13th 2018).
2. World Health Organization, *Noncommunicable Diseases Progress Monitor 2017*, 2017, ISBN: 978-92-4-151302-9.
3. Food and Agriculture Organization of the United Nations, World Health Organization, University UN, Human energy requirements, Food and Agriculture Organization of the United Nations, Rome, 2001.
4. G. Cediel, D. Gaitan-Charry, L. I. González-Zapata, E. Pérez-Tamayo and O. Sarmiento Malnutrition in all its forms and socioeconomic status in Colombia, in *Annals of Nutrition And Metabolism*, ed. M. Carrera, A. Gil and A. Martinez, IUNS 21st International Congress of Nutrition, Buenos Aires, Argentina, October 15–20 2017, p. 240.
5. M. Miranda, A. Bento and A. M. Aguilar Malnutrition in all its forms and socioeconomic status in Bolivia, in *Annals of Nutrition And Metabolism*, ed. M. Carrera, A. Gil and A. Martinez, IUNS 21st International Congress of Nutrition, Buenos Aires, Argentina, October 15–20 2017, p. 239.
6. K. Quinto, López and D. de Romaña, Malnutrition in all its forms and socioeconomic status in Peru, in *Annals of Nutrition And Metabolism*, ed. M. Carrera, A. Gil and A. Martinez, IUNS 21st International Congress of Nutrition. Buenos Aires, Argentina, October 15–20 2017, p. 242.
7. Ministerio de Salud, Indicadores Básicos de Salud Chile, (DEIS) DdEeIeS, 2015.
8. M. Mazariegos, M. F. Kroker-Lobos and M. Ramirez-Zea Malnutrition in all its forms and socioeconomic indicators in Guatemala, in *Annals of Nutrition And Metabolism*, ed. M. Carrera, A. Gil and A. Martinez, IUNS 21st International Congress of Nutrition, Buenos Aires, Argentina, October 15–20 2017, p. 241.
9. C. Severi, M. Medina, X. Moratorio, M. V. Natero Felipe and P. Barreto, Malnutrition in all its forms and socioeconomic status in Uruguay, in *Annals of Nutrition And Metabolism*, ed. M. Carrera, A. Gil and A. Martinez, IUNS 21st International Congress of Nutrition, Buenos Aires, Argentina, October 15–20 2017, p. 243.
10. D. Canella, A. C. Duran and C. Monteiro, Malnutrition in all its forms and socioeconomic status in Brazil, in *Annals of Nutrition And Metabolism*, ed. M. Carrera, A. Gil and A. Martinez, IUNS 21st International Congress of Nutrition, Buenos Aires, Argentina, October 15–20 2017, p. 239.
11. M. J. Ramirez-Luzuriaga, P. Belmont, W. F. Waters and W. Freire, Malnutrition in all its forms differs by socioeconomic status, education level and race/ethnicity in Ecuador, in *Annals of Nutrition And Metabolism*, ed. M. Carrera, A. Gil and A. Martinez, IUNS 21st

International Congress of Nutrition, Buenos Aires, Argentina, October 15–20 2017, p. 241.

12. C. Batis, E. Denova-Gutiérrez, B. Estrada-Velasco and J. Á. Rivera-Dommarco, Malnutrition in all its forms and socioeconomic status in Mexico, in *Annals of Nutrition And Metabolism*, ed. M. Carrera, A. Gil and A. Martinez, IUNS 21st International Congress of Nutrition, Buenos Aires, Argentina, October 15–20 2017, p. 242.
13. D. Navarro-Rosenblatt, M. F. Mujica, S. Lopez-Arana and C. Corvalan, Malnutrition in all its forms and socioeconomic status in Chile, in *Annals of Nutrition And Metabolism*, ed. M. Carrera, A. Gil and A. Martinez, IUNS 21st International Congress of Nutrition, Buenos Aires, Argentina, October 15–20 2017, p. 243.
14. M. E. Zapata, A. I. Soruco and E. Carmuega, Malnutrition in all its forms and socioeconomic status in Argentina, in *Annals of Nutrition And Metabolism*, ed. M. Carrera, A. Gil and A. Martinez, IUNS 21st International Congress of Nutrition, Buenos Aires, Argentina, October 15–20 2017, p. 238.
15. GBD 2016 Mortality Collaborators, Global, regional, and national under-5 mortality, adult mortality, age-specific mortality, and life expectancy, 1970-2016: a systematic analysis for the Global Burden of Disease Study 2016, *Lancet*, 2017, **390**(10100), 1084–1150.
16. World Health Organization, *Diet, nutrition and the prevention of chronic diseases: report of a joint WHO/FAO expert*, World Health Organization, Geneva, 2003.
17. World Health Organization (WHO), Global strategy on diet, physical activity and health, France, 2004.
18. B. Swinburn, G. Sacks and S. Vandevijvere, *et al.*, INFORMAS (International Network for Food and Obesity/non-communicable diseases Research, Monitoring and Action Support): overview and key principles, *Obes. Rev.*, 2013, **14**(Suppl 1), 1–12.
19. K. Giskes, F. van Lenthe, M. Avendano-Pabon and J. Brug, A systematic review of environmental factors and obesogenic dietary intakes among adults: are we getting closer to understanding obesogenic enviroments?, *Obes. Rev.*, 2011, **12**, e95–e106.
20. S. Barquera, L. Hernández-Barrera, S. J. Rothenberg and E. Cifuentes, The obesogenic environment around elementary schools: food and beverage marketing to children in two Mexican cities, *BMC Public Health*, 2018, **18**(1), 461.
21. P. B. Ford and D. Dzewaltowski, Disparities in obesity prevalence due to variation in the retail food environment: three testable hypotheses, *Nutr. Rev.*, 2008, **66**(4), 216–228.
22. M. L. Costa Louzada, A. P. Martins and D. S. Canella, *et al.*, Ultra-processed foods and the nutritional dietary profile in Brazil, *Rev. Saude. Publica.*, 2015, **49**, 38.
23. J. A. Marrón-Ponce and T. G. Sánchez-Pimienta, Louzada MLDC, Batis C. Energy contribution of NOVA food groups and sociodemographic

determinants of ultra-processed food consumption in the Mexican population, *Public Health Nutr.*, 2018, **21**(1), 87–93.
24. B. Cornwell, E. Villamor, M. Mora-Plazas, C. Marin, C. A. Monteiro and A. Baylin, Processed and ultra-processed foods are associated with lower-quality nutrient profiles in children from Colombia, *Public Health Nutr.*, 2018, **21**(1), 142–147.
25. PAHO, *Ultra-processed food and drink products in Latin America: Trends, impact on obesity, policy implications*, Pan American Health Organization, Washington, DC, 2015.
26. C. A. Monteiro, G. Cannon and R. Levy, *et al.*, NOVA. The star shines bright, *World Nutrition*, 2016, **7**(1–3), 10.
27. G. Cediel, M. Reyes and M. L. da Costa Louzada, *et al.*, Ultra-processed foods and added sugars in the Chilean diet (2010), *Public Health Nutr.*, 2017, 1–9.
28. M. L. Louzada, A. P. Martins and D. S. Canella, *et al.*, Impact of ultra-processed foods on micronutrient content in the Brazilian diet, *Rev. Saude. Publica.*, 2015, **49**, 45.
29. D. S. Canella, R. B. Levy and A. P. Martins, *et al.*, Ultra-processed food products and obesity in Brazilian households (2008-2009), *PLoS One*, 2014, **9**(3), e92752.
30. R. D. Mendonça, A. C. Lopes, A. M. Pimenta, A. Gea, M. A. Martinez-Gonzalez and M. Bes-Rastrollo, Ultra-Processed Food Consumption and the Incidence of Hypertension in a Mediterranean Cohort: The Seguimiento Universidad de Navarra Project, *Am. J. Hypertens.*, 2017, **30**(4), 358–366.
31. E. Gourd, Ultra-processed foods might increase cancer risk, *Lancet Oncol.*, 2018, **19**(4), e186.
32. M. A. Colchero, B. M. Popkin, J. A. Rivera and S. W. Ng, Beverage purchases from stores in Mexico under the excise tax on sugar sweetened beverages: observational study, *BMJ*, 2016, **352**, h6704.
33. M. A. Colchero, J. A. Zavala, C. Batis, T. Shamah-Levy and J. A. Rivera-Dommarco, Changes in prices of taxed sugar-sweetened beverages and nonessential energy dense food in rural and semi-rural areas in Mexico, *Salud. Publica. Mex.*, 2017, **59**(2), 137–146.
34. Gobierno de Brasil. Lei de Alimentação Escolar N°11.947. Brasilia; 2009.
35. Food and Agriculture Organization of the United Nations (FAO), Pan American Health Organization (PAHO), Approval of a New Food Act in Chile: Process Summary. Entry into force: June 2016, Santiago de Chile, 2017.
36. E. V. Kleef and H. Dagevos, The growing role of front-of-pack nutrition profile labeling: a consumer perspective on key issues and controversies, *Crit. Rev. Food Sci. Nutr.*, 2015, **55**(3), 291–303.
37. A. J. Roodenburg, B. M. Popkin and J. C. Seidell, Development of international criteria for a front of package food labelling system: the International Choices Programme, *Eur. J. Clin. Nutr.*, 2011, **65**(11), 1190 1200.

38. World Cancer Research Fund International, Our policy framework to promote healthy diets & reduce obesity, 2015. https://www.wcrf.org/int/policy/nourishing/our-policy-framework-promote-healthy-diets-reduce-obesity (accessed March 14th 2018).
39. C. A. Roberto, M. Shivaram, O. Martinez, C. Boles, J. L. Harris and K. D. Brownell, The Smart Choices front-of-package nutrition label. Influence on perceptions and intake of cereal, *Appetite*, 2012, **58**(2), 651–657.
40. C. N. Mhurchu, H. Eyles and Y. H. Choi, Effects of a Voluntary Front-of-Pack Nutrition Labelling System on Packaged Food Reformulation: The Health Star Rating System in New Zealand, *Nutrients*, 2017, **9**(8), 918.
41. K. L. Hawley, C. A. Roberto, M. A. Bragg, P. J. Liu, M. B. Schwartz and K. D. Brownell, The science on front-of-package food labels, *Public Health Nutr.*, 2013, **16**(3), 430–439.
42. (FAO) FaAOotUN, WHO. CODEX ALIMENTARIUS International Food Standards. Guidelines on Nutrition Labelling CAC/GL 2-1985, 1985.
43. Consejo de Ministros de Integración Económica Centroamericana (COMIECO), Resolución N° 281-2012 (COMIECO LXII) Requisitos para etiquetado nutricional de productos alimenticios preenvasados para consumo humano para la población a partir de 3 años de edad. In: COMIECO, editor. Nicaragua: La Gaceta Diario Oficial, 2012. p. 8.
44. Mercado Común del Sur (MERCOSUR), Reglamento Técnico MERCOSUR para Rotulación de Alimentos Envasados. In: Ministerio de Salud Pública MdRE, Ministerio de Economía y Finanzas, Ministerio de Industria EyM, Ministerio de Ganadería AyP, editors. Montevideo: IM.P.O Diario Oficial, 2006. p. 22.
45. Ministerio de la Producción, Decreto legislativo que aprueba la Ley de etiquetado y verificación de los reglamentos técnicos de los productos industriales manufacturados In: Producción Mdl, editor. 610509, Lima, El Peruano, 2016, p. 2.
46. Ministerio de la Protección Social, Reglamento técnico sobre los requisitos de rotulado o etiquetado nutricional que deben cumplir los alimentos envasados para consumo humano. In: Social MdlP, editor. Colombia: Ministerio de la Protección Social, 2011. p. 56.
47. Ministerio de Salud (MINSAL), Modifica Decreto Supremo N°977, de 1996, Reglamento Sanitario de los Alimentos. In: Salud Md, editor. IdDO 918813: Diario Oficial de la República de Chile, 2015, p. 5.
48. Secretaría de Economía Estados Unidos Mexicanos, NORMA OFICIAL MEXICANA NOM-051-SCFI/SSA1-2010, Especificaciones generales de etiquetado para alimentos y bebidas no alcohólicas preenvasados-Información comercial y sanitaria, México, D.F., 2010.
49. K. G. Grunert, J. M. Wills and L. Fernández-Celemín, Nutrition knowledge, and use and understanding of nutrition information on food labels among consumers in the UK, *Appetite*, 2010, **55**(2), 177–189.

50. K. G. Grunert, L. Fernández-Celemín, J. M. Wills, S. Storcksdieck Genannt Bonsmann and L. Nureeva, Use and understanding of nutrition information on food labels in six European countries, *Z Gesundh Wiss*, 2010, **18**(3), 261–277.
51. Food and Agriculture Organization of the United Nations (FAO) and World Health Organization (WHO), CODEX ALIMENTARIUS COMMISSION Discussion paper on consideration of issues regarding Front-of-Pack nutrition labelling CX/FL 17/44/7, Asunción, Paraguay, 2017.
52. Chilean Ministry of Health, Modifica Decreto Supremo N° 977, de 1996, Reglamento Sanitario de los Alimentos, 2015.
53. Ministerio de Salud Pública de Ecuador, Reglamento de etiquetado de alimentos procesados para consumo humano. In: Pública MdS, editor., 2014, p. 11.
54. E. van Kleef, H. van Trijp, F. Paeps and L. Fernández-Celemín, Consumer preferences for front-of-pack calories labelling, *Public Health Nutr.*, 2008, **11**(2), 203–213.
55. R. Endevelt, I. Grotto, R. Sheffer, *et al.*, Policy and practice - Regulatory measures to improve the built nutrition environment for prevention of obesity and related morbidity in Israel, World Health Organization, 2017.
56. T. E. Emrich, Y. Qi, J. E. Mendoza, W. Lou, J. E. Cohen and M. R. L'abbé, Consumer perceptions of the Nutrition Facts table and front-of-pack nutrition rating systems, *Appl. Physiol., Nutr., Metab.*, 2014, **39**(4), 417–424.
57. E. van Herpen and H. C. Trijp, Front-of-pack nutrition labels. Their effect on attention and choices when consumers have varying goals and time constraints, *Appetite*, 2011, **57**(1), 148–160.
58. N. Savoie, K. Barlow Gale, K. L. Harvey, M. A. Binnie and L. Pasut, Consumer perceptions of front-of-package labelling systems and healthiness of foods, *Can. J. Public Health*, 2013, **104**(5), e359–e363.
59. B. Kelly, C. Hughes and K. Chapman, *et al.*, Consumer testing of the acceptability and effectiveness of front-of-pack food labelling systems for the Australian grocery market, *Health Promot. Int.*, 2009, **24**(2), 120–129.
60. S. Goodman, D. Hammond, R. Hanning and J. Sheeshka, The impact of adding front-of-package sodium content labels to grocery products: an experimental study, *Public Health Nutr.*, 2013, **16**(3), 383–391.
61. P. Ducrot, C. Méjean and C. Julia, *et al.*, Effectiveness of Front-Of-Pack Nutrition Labels in French Adults: Results from the NutriNet-Santé Cohort Study, *PLoS One*, 2015, **10**(10), e0140898.
62. T. E. Emrich, J. E. Mendoza and M. R. L'Abbé, Effectiveness of front-of-pack nutrition symbols: a pilot study with consumers, *Can. J. Diet Pract. Res.*, 2012, **73**(4), 200–203.
63. A. Arrúa, L. Machín and M. R. Curutchet, *et al.*, Warnings as a directive front-of-pack nutrition labelling scheme: comparison with the

Guideline Daily Amount and traffic-light systems, *Public Health Nutr.*, 2017, **20**(13), 2308–2317.
64. L. Machín, J. Aschemann-Witzel, M. R. Curutchet, A. Giménez and G. Ares, Traffic Light System Can Increase Healthfulness Perception: Implications for Policy Making, *J. Nutr. Educ. Behav.*, 2018, **50**(7), 668–674.
65. T. Bollard, N. Maubach, N. Walker and C. N. Mhurchu, Effects of plain packaging, warning labels, and taxes on young people's predicted sugar-sweetened beverage preferences: an experimental study, *Int. J. Behav. Nutr. Phys. Act.*, 2016, **13**(1), 95.
66. A. Arrúa, M. R. Curutchet and N. Rey, *et al.*, Impact of front-of-pack nutrition information and label design on children's choice of two snack foods: Comparison of warnings and the traffic-light system, *Appetite*, 2017, **116**, 139–146.
67. N. Khandpur, P. de Morais Sato and L. A. Mais, *et al.*, Are Front-of-Package Warning Labels More Effective at Communicating Nutrition Information than Traffic-Light Labels? A Randomized Controlled Experiment in a Brazilian Sample, *Nutrients*, 2018, **10**, 6.
68. Agency SF, A qualitative study concerning the Keyhole's influence over 25 years on product development, Uppsala, Sweden, 2015.
69. National Food Agency Sweden, Regulations amending the National Food Agency's Regulations (SLVFS 2005:9) on the use of a particular symbol, Sweden, 2015, p. 17.
70. R. K. Johnson, A. H. Lichtenstein and P. M. Kris-Etherton, *et al.*, Enhanced and Updated American Heart Association Heart-Check Front-of-Package Symbol: Efforts to Help Consumers Identify Healthier Food Choices, *J. Acad. Nutr. Diet.*, 2015, **115**(6), 876–884.
71. Food and Drink Federation (FDF), Science behind Guideline Daily Amounts, 2009, p. 2.
72. Institute of Grocery Distribution (IGD), Report of the IGD/PIC Industry Nutrition Strategy Group Technical Working Group on Guideline Daily Amounts (GDAs), Watford, UK, 2005.
73. F. AlimentumImplantación del sistema GDA'S/CDO en el sector de alimentación y bebidas español, Ventajas para empresas y consumidores, Madrid, España, 2009.
74. Unilever, Guideline daily amount, p. 3.
75. Food Standards Agency, Guide to creating a front of pack (FoP) nutrition label for pre-packed products sold through retail outlets, UK, 2016.
76. Food Standards Agency, INT 13/03/01 Front of Pack Nutrition Labelling - Proposals of the new UK-wide scheme, 2013, p. 9.
77. C. Hodgkins, J. Barnett and G. Wasowicz-Kirylo, *et al.*, Understanding how consumers categorise nutritional labels: a consumer derived typology for front-of-pack nutrition labelling, *Appetite*, 2012, **59**(3), 806–817.
78. M. W. Becker, R. P. Sundar, N. Bello, R. Alzahabi, L. Weatherspoon and L. Bix, Assessing attentional prioritization of front-of-pack nutrition labels using change detection, *Appl. Ergon.*, 2016, **54**, 90–99.

79. D. J. Graham, C. Heidrick and K. Hodgin, Nutrition Label Viewing during a Food-Selection Task: Front-of-Package Labels vs Nutrition Facts Labels, *J. Acad. Nutr. Diet.*, 2015, **115**(10), 1636–1646.
80. H. J. Yoo, L. Machín and A. Arrúa, *et al.*, Children and adolescents' attitudes towards sugar reduction in dairy products, *Food Res. Int.*, 2017, **94**, 108–114.
81. C. Julia, S. Péneau and C. Buscail, *et al.*, Perception of different formats of front-of-pack nutrition labels according to sociodemographic, lifestyle and dietary factors in a French population: cross-sectional study among the NutriNet-Santé cohort participants, *BMJ Open*, 2017, **7**(6), e016108.
82. P. J. Kenny, Reward mechanisms in obesity: new insights and future directions, *Neuron*, 2011, **69**(4), 664–679.
83. S. M. Nicola, Reassessing wanting and liking in the study of mesolimbic influence on food intake, *Am. J. Physiol.: Regul., Integr. Comp. Physiol.*, 2016, **311**(5), R811–R840.
84. G. J. Wang, N. D. Volkow, P. K. Thanos and J. S. Fowler, Similarity between obesity and drug addiction as assessed by neurofunctional imaging: a concept review, *J. Addict. Dis.*, 2004, **23**(3), 39–53.
85. W. Hoyer, An Examination of consumer decision making for a Common Repeat Purchase Product, *J. Consum. Res.*, 1984, **11**(3), 822–829.
86. C. A. Roberto, J. Baik, J. L. Harris and K. D. Brownell, Influence of licensed characters on children's taste and snack preferences, *Pediatrics*, 2010, **126**(1), 88–93.
87. P. Letona, V. Chacon, C. Roberto and J. Barnoya, A qualitative study of children's snack food packaging perceptions and preferences, *BMC Public Health*, 2014, **14**, 1274.
88. M. A. Lapierre, S. E. Vaala and D. L. Linebarger, Influence of licensed spokescharacters and health cues on children's ratings of cereal taste, *Arch. Pediatr. Adolesc. Med.*, 2011, **165**(3), 229–234.
89. B. Sütterlin and M. Siegrist, Simply adding the word "fruit" makes sugar healthier: The misleading effect of symbolic information on the perceived healthiness of food, *Appetite*, 2015, **95**, 252–261.
90. F. Grabenhorst, F. P. Schulte, S. Maderwald and M. Brand, Food labels promote healthy choices by a decision bias in the amygdala, *NeuroImage*, 2013, **74**, 152–163.
91. P. Ducrot, C. Méjean and C. Julia, *et al.*, Objective Understanding of Front-of-Package Nutrition Labels among Nutritionally At-Risk Individuals, *Nutrients*, 2015, **7**(8), 7106–7125.
92. I. Borgmeier and J. Westenhoefer, Impact of different food label formats on healthiness evaluation and food choice of consumers: a randomized-controlled study, *BMC Public Health*, 2009, **9**, 184.
93. L. Machín, M. Cabrera, M. R. Curutchet, J. Martínez, A. Giménez and G. Ares, Consumer Perception of the Healthfulness of Ultra-processed Products Featuring Different Front-of-Pack Nutrition Labeling Schemes, *J. Nutr. Educ. Behav.*, 2017, **49**(4), 330–338, e1.

94. N. Babio, P. Vicent, L. López, A. Benito, J. Basulto and J. Salas-Salvadó, Adolescents' ability to select healthy food using two different front-of-pack food labels: a cross-over study, *Public Health Nutr.*, 2014, **17**(6), 1403–1409.
95. C. A. Roberto, M. A. Bragg and M. B. Schwartz, *et al.*, Facts up front versus traffic light food labels: a randomized controlled trial, *Am. J. Prev. Med.*, 2012, **43**(2), 134–141.
96. N. Babio, L. López and J. Salas-Salvadó, Capacity analysis of health food choice by reference to consumers in two models of nutritional labeling; crossover study, *Nutr. Hosp.*, 2013, **28**(1), 173–181.
97. N. Maubach, J. Hoek and D. Mather, Interpretive front-of-pack nutrition labels. Comparing competing recommendations, *Appetite*, 2014, **82**, 67–77.
98. R. Pieters and M. Wedel, Attention Capture and Transfer in Advertising: Brand, Pictorial, and Text-Size Effects, *J. Marketing*, 2004, **68**(2), 36–50.
99. M. Cabrera, L. Machín and A. Arrúa, *et al.*, Nutrition warnings as front-of-pack labels: influence of design features on healthfulness perception and attentional capture, *Public Health Nutr.*, 2017, **20**(18), 3360–3371.
100. Gobierno de México Secretaría de Salud Pública, Nuevo Etiquetado Frontal Nutrimental, 2015. https://www.gob.mx/salud/en/documentos/nuevo-etiquetado-frontal-nutrimental (accessed March 14th 2018).
101. Comisión Federal para la Protección de Riesgos Sanitarios (COFEPRIS), Manual de Etiquetado Frontal Nutrimental, 2016.
102. L. Tolentino-Mayo, S. Rincón-Gallardo Patiño, L. Bahena-Espina, V. Ríos and S. Barquera, Conocimiento y uso del etiquetado nutrimental de alimentos y bebidas industrializados en México, *Salud. Publica. Mex.*, 2018, **60**, 328–337.
103. Alianza por la Salud Alimentaria, El etiquetado frontal nutrimental de bebidas y alimentos en México. A dos años de su implementación, 2016, p. 1–4.
104. D. Stern, L. Tolentino and S. Barquera, Revisión del etiquetado frontal: análisis de las Guías Diarias de Alimentación (GDA) y su comprensión por estudiantes de nutrición en México, Instituto Nacional de Salud Pública de México, Cuernavaca, 2011.
105. A. Díaz, P. Veliz and G. Rivas-Mariño, *et al.*, Etiquetado de alimentos en Ecuador: implementación, resultados y acciones pendientes, *Rev. Panam. Salud. Públ.*, 2017, **41**, e54.
106. Organización Panamericana de la Salud (OPS), Recomendaciones de la Consulta de Expertos de la Organización Panamericana de la Salud sobre la promoción y publicidad de alimentos y bebidas no alcohólicas dirigidas a los niños en la Región de las Américas, Washington, D.C, 2011.
107. W. B. Freire, W. F. Waters, G. Rivas-Mariño, T. Nguyen and P. Rivas, A qualitative study of consumer perceptions and use of traffic light food labelling in Ecuador, *Public Health Nutr.*, 2017, **20**(5), 805–813.

108. Ministerio de Salud (MINSAL). Ley de Alimentos, Nuevo etiquetado de alimentos, 2016. http://www.minsal.cl/ley-de-alimentos-nuevo-etiquetado-de-alimentos/ (accessed March 14 2018).
109. C. Corvalan, M. Reyes, M. L. Garmendia and R. Uauy, Structural responses to the obesity and non-communicable diseases epidemic: Update on the Chilean law of food labelling and advertising, *Obes. Rev.*, 2019, **20**, 367–374.
110. Ministerio de Salud de Chile, Informe de Evaluación de la Implementación de la Ley sobre composición nutricional de los alimentos y su publicidad, Chile, 2017.
111. Ministerio de Salud de Chile, Informe de Evaluación de la implementación de la Ley sobre composición nutricional de los alimentos y su publicidad, Ministerio de Salud, Santiago, Chile, 2018.
112. Ministerio de Salud de Chile Instituto de la Comunicación e Imagen Universidad de Chile Demoscopica, Informe de Resultados: Descripción de las percepciones y actitudes de los/asconsumidores respecto a las medidas estatales en el marco de la implementación del Decreto 13/15, Chile, 2017.
113. S. Olivares, B. Leyton, N. Bustos, *et al.*, Actitudes de escolares chilenos de distinto nivel socioeconómico al inicio de la implementación de la ley que regula la venta y publicidad de alimentos altos en nutrientes críticos, *Nutr. Hosp.*, 2017, DOI: 10.20960/nh.499.
114. T. Correa, C. Fierro, M. Reyes, F. R. Dillman Carpentier, L. S. Taillie and C. Corvalan, *Int. J. Behav. Nutr. Phys. Act.*, 2019, **16**, 21.
115. Ministerio de Salud Pública de Uruguay, Proyecto de decreto del Ministerio de Salud Pública: Rotulado de Alimentos Envasados en Uruguay, Montevideo, Uruguay; 2017.
116. Plataforma de Seguridad Alimentaria y Nutricional (SAN), Ley de promoción de la alimentación saludable para niños, niñas y adolescentes, 2018. http://plataformacelac.org/ley/15.
117. Pan American Health Organization (PAHO), Pan American Health Organization Nutrient Profile Model, Wahington, DC, 2016.

Subject Index

advisory warning label (AWL), 134, 135
alcoholic drinks, 8–9
alcopop, 8
allergen declarations, 116–118
allergen-free food regulation and labelling, 120–123
allergen-free statements, 158
allergenic foods, 109, 112–114, 117
allergen labelling lists, 115
anaphylaxis, 107, 127, 132, 158
antioxidant claims, 171
Article 13.1 health claims, 18, 65–67, 68, 80
Article 13.5 health claims, 18, 22, 65, 67, 69, 80
Article 14 health claims, 67–68
Article 14a health claims, 67, 68, 80
Article 14b health claims, 68, 80
aspartame, 142, 143
Australian and New Zealand star system, 7
AWL *see* advisory warning label

BACCHUS Best Practice Guide, 75

calcium, 68, 167
Canadian Food Inspection Agency (CFIA), 154
Canadian food regulators, 154
CAP *see* Common Agricultural Policy
CCFL *see* Codex Committee on Food Labelling
cereal consumption, 103
CFIA *see* Canadian Food Inspection Agency
childhood development and health claims, 18, 82
childhood obesity, 3
claim options, FSANZ, 166–171
    health claims, 167–169
    nutrient content claims, 167
CLYMBOL, 73, 81, 82, 84, 85, 87, 88, 89, 91, 94, 96, 101, 105
CODEX Alimentarius Commission, 4–5, 29, 108, 109, 142, 150, 155, 159, 182
Codex Committee on Food Labelling (CCFL), 5, 154
CODEX list, food allergens, 110, 111, 115
CODEX STAN 1-1985, 4
CODEX Standard 118-1979, 121
CODEX Standard 188-1979, 121, 122
coeliac disease, 157
Common Agricultural Policy (CAP), 8
Community Register of Claims, 70
consumer education, 51
consumer organisations, 1
consumer understanding test (CUT method), 97–98, 104
Council Directive 77/94/EEC, 108
CUT *see* consumer understanding test

DEFRA *see* Department for Environment, Food & Rural Affairs
Department for Environment, Food & Rural Affairs (DEFRA), 44
discretionary food labelling measures

gluten-free/allergen-free claims, 157–158
health claims, 155–157
with regulatory requirements, 155–158
disease risk claims reduction, 82
docosahexaenoic acid, 68
Dutch keyhole system, 7

ECJ *see* European Court of Justice
EFSA *see* European Food Safety Authority
electrodermal reactivity, 100–101
ELISA *see* enzyme-linked immunosorbent assay methods
energy content, alcoholic drink, 9
energy reduction, nutrition label, 64
enzyme-linked immunosorbent assay (ELISA) methods, 119, 123, 133
estimated safe and adequate daily dietary intakes (ESADDI), 167
EU FP6 EuroFIR, 46
EU member states (MS), 63, 76, 81
EU member state laws collection, 31–32
nutrition labelling issues, national authorities, 32
EUR-Lex portal, 30
European Court of Justice (ECJ), 16, 23
European Court of Justice (ECJ) (C-19/15), 17
European Food Safety Authority (EFSA), 15, 19, 20, 21, 23, 64, 67, 69, 70, 76
European Union, food law, 30–32
EUR-Lex portal, 30
nutrition labelling issues, Official European Commission Portal, 31
eye-tracking study, 100, 104–105

Fair Trading Act, 1986, 165
FALCPA *see* Food Allergen Labelling and Consumer Protection Act
FBOs *see* food business operators
FCPC *see* Food Consumer Products of Canada
FDNC *see* Food Databanks National Capability
FIC *see* food information to consumers
FLABEL study, 86
flavoured teas, 5
Food Allergen Labelling and Consumer Protection Act (FALCPA), 110, 111, 115, 116, 122, 153
food allergen labelling regulation
allergen declarations, 116–118
mandatory allergen labelling, 108–110
regulated priority food allergens, 110–116
voluntary precautionary allergen labelling, 118–119
food-allergic adolescents, 132
food allergy prevalence, Canada, 152
food business operators (FBOs), 29, 30, 46–48, 56, 95, 115, 116, 117, 119, 131, 133, 134, 165, 166, 167, 169, 174, 176
Food Consumer Products of Canada (FCPC), 148
Food Databanks National Capability (FDNC), 50
food information to consumers (FIC), 28, 29, 30, 31, 50, 51
food law
access to authoritative sources, 30–32
general requirements, 133
food regulators, 158
Food Safety Authority of Ireland (FSAI), 43
food safety policy, 14–16
business operators, 14
general food law, 14

Food Standards Australia New Zealand (FSANZ), 111, 121, 164, 165
antioxidant claims, 171
branded products, developing, 172–173
claim options, navigating, 166–171
legislation, 115
self-substantiation dossier, developing, 173–174
vitalvegetables®, 172–173
Food Standards Australia New Zealand's Nutrient Profiling Scoring Criterion (FSANZ NPSC), 84, 89
FOP *see* front-of-package labelling
front-of-package (FOP) labelling, 7–8, 43, 183
Brazil, 188
Central American and Caribbean countries, 188
Colombia, 188
Latin American countries, 183–188
Mexico, Ecuador and Chile, 184–187
systems, 149
Uruguay and Peru, 188
front-of-pack nutrition labelling information, 148–150
Canada's proposed mandatory labelling requirements, 149–150
health and policy rationale, 148–149
FSA *see* UK Food Standards Agency
FSAI *see* Food Safety Authority of Ireland
FSANZ *see* Food Standards Australia New Zealand
FSANZ NPSC *see* Food Standards Australia New Zealand's Nutrient Profiling Scoring Criterion
function health claims, 18
GDAs *see* guideline daily amounts
general (non-specific) health claims, 80, 89
general level health claims, 167
general non-specific health benefits, 68
glutelins, 157
gluten, 120, 121, 122
gluten-free/allergen-free claims, 157–158
guideline daily amounts (GDAs), 183
guidelines, government, 32–44
e-learning courses, 43–44
EuroFIR guideline, recipe calculation, 46–48, 49
FIC implementation, 42
Lithuanian recipe calculation, 48
manufacturers' analyses, 45
non-specialist user *vs.* expert assisted recipe calculation, 48, 50
nutrition legislation, comprehensive information, 42
on obtaining nutrient values, 44–50
online guidelines, 43–44
question and answer (Q&A) format, 32, 42
values calculation, 45–50
voluntary labelling, additional guidelines, 43

halo effect, 97
Health Canada, 154, 156, 158
health claims, 65, 80
Article 13.1, 18, 65–67, 68, 80
Article 13.5, 18, 22, 65, 67, 69, 80
Article 14, 67–68
breakfast study, 102–103
consumer purchases, testing methods, 99–101
consumer understanding of, 73–74

consumer understanding test (CUT method), 97–98
consumption effects, testing methods, 102–103
effects, feasibility, 96
electrodermal reactivity study, 100–101
eye-tracking study, 100
hard laddering (online), 98–99
implicit attitude study, 101
labelling of foods, 87–89
reduction, 82
sales data, 100
soft laddering interviews (face-to-face), 98
testing methods, consumer understanding, 96–99
theory-driven criteria, 95
validity and reliability, general criteria, 96

health-related quality of life (HRQL), 128, 131

health symbol schemes
breakfast study, 102–103
consumer purchases, testing methods, 99–101
consumer understanding test (CUT method), 97–98
consumption effects, testing methods, 102–103
effects, feasibility, 96
electrodermal reactivity study, 100–101
eye-tracking study, 100
hard laddering (online), 98–99
implicit attitude study, 101
sales data, 100
soft laddering interviews (face-to-face), 98
testing methods, consumer understanding, 96–99
theory-driven criteria, 95
validity and reliability, general criteria, 96

high in fibre, nutrition labels, 64

high in PUFA, nutrition labels, 65

high-level health claims, 167

Hordeum, 122

HRQL *see* health-related quality of life

hypersensitivity, 109

iFAAM *see* Integrated Approaches to Food Allergen and Allergy Risk Management

ILT *see* instructor-led training courses

implicit attitude, 101

individual (micro) system, 129–132

informed consumer choices, 84–85

instructor-led training (ILT) courses, 52–55

Integrated Approaches to Food Allergen and Allergy Risk Management (iFAAM), 135

interpretive labelling, 7

Japanese regulatory system, 115

label, nutrients, 1

labelling requirements, 5–6

Latin American countries
FOP labelling in, 183–188
undernutrition, 180

low in fat, nutrition label, 64

"magic bullet" effect, 97

mandatory food allergen labelling measures, 152–155
allergen precautionary labelling, 154–155
as ingredients list part, 153–154

measurement uncertainty, 10, 11

mechanically tenderised beef (MTB), 151

mechanically tenderised meat products
Canadian requirements, 151
mandatory labelling of, 150–152
US requirements, 152

mental noise, 130
MTB *see* mechanically tenderised beef

NCDs *see* non-communicable diseases
New Zealand enforcement agency, 176
New Zealand Food Act, 2014, 165
New Zealand Food Composition Database, 169–171
NFP *see* Nutrition Fact Panels
NFt *see* Nutrition Facts table (Canada)
NHCR *see* Nutrition and Health Claims Regulation
NLEA *see* Nutrition Labelling Education Act
NLS *see* nutritional labelling software
non-communicable diseases (NCDs), 143, 144, 148, 180
non-directive GDA, 184
Nordic Keyhole scheme, 43
North American food regulators, 144
North American jurisdictions
  discretionary labelling decisions, 142
  mandatory food allergen labelling measures, 152–155
  mandatory labelling requirements, 142
  mechanically tenderised meat products, 150–152
  nutrition facts labelling, 143–150
  risk management labelling, 142–155
novel foods legislation, 71–73
NPSC *see* nutrient profiling scoring criteria
NQA *see* nutrient quality assurance
NRV *see* nutrient reference value
nutrient profiles, 16, 17, 68
  issue, 21–22
nutrient profiling scoring criteria (NPSC), 169
nutrient quality assurance (NQA), 173
nutrient reference value (NRV), 65
nutri-score scheme, 43
nutritional declarations, 2, 4, 5, 8, 44
  labelling with, 86
nutritional labelling software (NLS), 46
nutritional poverty, 3
nutritional quality, 83–84
Nutrition and Health Claims Regulation (NHCR), 14–16
  application, 20
  Article 2, 18
  Article 8, 18
  authoritative statement of, 19–22
  claims, 16–17
  consumers protection, 15–16
  food labelling, 17
  free movement of foods, 15
  generic (non-specific) claims, 19
  harmonisation of, 14–16
  health claims, 18–19
  labelling commercial aspects, 16–17
  markets stimulation, 15–16
  nutrient profiles issue, 21–22
  nutrition claims, 18
  permitted health claims, 22
  permitted nutrition claims, 22
  regulation adoption, 15
  risk assessment to risk management, 20–21
  scientific substantiation, 19
  suspension claims, 22–23
Nutrition and Health Claims Regulations, 2
nutrition claims, 64–65, 80
  labelling of foods, 87
  penetration of, 82
Nutrition Fact Panels (NFP), 149
Nutrition Facts label (USA), 144

Nutrition facts labelling, North America, 143–150
Canadian nutrition facts table, 145, 146
front-of-pack nutrition labelling information, 148–150
US nutrition facts label, 145, 147–148
Nutrition Facts table (NFt) (Canada), 144
nutrition information, 2, 3–4
nutrition labelling, 2, 4, 5
alcoholic drinks and, 8–9
Nutrition Labelling Education Act (NLEA), 141

oat beta-glucan, 68
obesity, 180
Official Journal (OJ), EUR-Lex portal, 30

PAHO *see* Pan American Health Organization report
PAL *see* precautionary allergen labelling
Pan American Health Organization (PAHO) report, 180, 181, 185
PARNUTS *see* particular nutritional uses
particular nutritional uses (PARNUTS), 108
phenylalanine, 142, 143
phenylketonuria (PKU), 142, 143
phenyl pyruvic acid, 143
PKU *see* phenylketonuria
Plant & Food Research, 174
key considerations, 172–173
plant stanol esters, 68
plant sterols, 68
precautionary allergen labelling (PAL), 118, 119, 123, 128
indirect impact of, 132
individual (micro) system, 129–132
multilevel perspective and, 134–135
risk management tool, 134
society (macro) system, 132–133
prepacked foods, 85
in EU MS, 85
health-related claims, 81–83
nutrition, prevalence of, 81–83
prolamins, 157
psychometric paradigm, 130

QRA *see* quantitative risk assessment
qualified health claim, 156
quality appraisal table, self-substantiation dossiers, 175
quantitative risk assessment (QRA), 134, 135

REDICLAIM *see* Reduction of disease risk claims on food and drinks
Reduction of disease risk claims on food and drinks (REDICLAIM), 75, 89
reference intake (RI), 65
REFIT *see* Regulatory Fitness and Performance program
regulation, FIC (1169/2011/EC), 4, 5
Regulation (EC) No 178/2002, 14
Article 14 of, 117
Regulation (EU) No 828/2014, 120
Regulation (EU) No 1169/2011, 28, 87, 91, 110, 111, 117, 120
Regulation (EC) No 1924/2006, 13, 14, 15, 23, 63, 64, 65, 67, 68, 73, 74, 75, 80, 83, 85, 87, 88, 89, 104
Article 1.2 of, 16
Article 10.3 of, 68
evaluation, process, 69–71
Regulation (EU) No 2015/2283, 23, 71, 73
Regulatory Fitness and Performance program (REFIT), 23, 83
response latencies (reaction times), 101
RI *see* reference intake
right to access, 13
risk perception, 130
risk reduction claims, 18

sales data, 100
scientific evidence, health claims, 18
scientific substantiation, 19
Secale, 122
self-substantiation dossiers, 173–174
quality appraisal table, 175
semi-directive traffic light system, 184
small and medium enterprises (SMEs), 74, 75, 133
small businesses, regulation impact, 74–75
SMEs *see* small and medium enterprises
society (macro) system, 132–133
soft laddering interviews (face-to-face), 98
source of vitamin C, nutrition label, 65
stakeholder methods, recommendations
for consumer scientists, 104–105
for (food) industry, 104
for regulators, 104
stress, 130
sugar-free, nutrition label, 64
supplementary nutrients, 5

tartrazine, 111
theory-driven criteria, 95
tolerances, 9–11
tonic, 9
traffic light labelling, 7–8
training courses, FIC, 50–51
trans unsaturated fatty acids (trans fats), 6–7
public consultation on, 6
Triticum, 122

UK Food Standards Agency (FSA), 44
undernutrition, Latin American countries, 180
United States Food and Drug Administration (US FDA) ESADDI values, 167
US Alcohol and Tobacco Tax and Trade Bureau, 116
USDA *see* US Department of Agriculture
USDA regulatory requirements, 152
US Department of Agriculture (USDA), 142, 152, 153
US FDA *see* US Food and Drug Administration
US Food and Drug Administration (US FDA), 116, 122, 142, 145, 153, 156, 167

VITAL™ *see* Voluntary Incidental Trace Allergen Labelling
vitalbones®, 172
vitalheart®, 172
vitalimmunity®, 172
vitalsight®, 172
vitalvegetables®, 172–173
vitamin C, 68
vitamin D, 18, 19
Voluntary Incidental Trace Allergen Labelling (VITAL™), 118

World Health Organization (WHO), 3, 4, 143, 180

"zero tolerance" policy, 120
Zespri®, 174